UNITED STATES DEPARTMENT OF THE INTERIOR
Oscar L. Chapman, Secretary

GEOLOGICAL SURVEY
W. E. Wrather, Director

Bulletin 955-D

GOLD PLACERS AND THEIR GEOLOGIC ENVIRONMENT IN NORTHWESTERN PARK COUNTY, COLORADO

BY

QUENTIN D. SINGEWALD

Prepared in cooperation with the
STATE OF COLORADO, THE COLORADO METAL MINING FUND
THE COLORADO MINING ASSOCIATION
AND THE COLORADO GEOLOGICAL SURVEY BOARD

Contributions to economic geology, 1947
(Pages 103-172)

UNITED STATES
GOVERNMENT PRINTING OFFICE
WASHINGTON : 1950

For sale by the Superintendent of Documents, U. S. Government Printing Office
Washington 25, D. C. - Price $1.25 cents

CONTENTS

ILLUSTRATIONS

GOLD PLACERS AND THEIR GEOLOGIC ENVIRONMENT IN NORTHWESTERN PARK COUNTY, COLORADO

By Quentin D. Singewald

ABSTRACT

The placer district of northwest Park County covers more than 250 square miles. The area embraces the mountain province of the northward-trending Mosquito Range and the northeastward-trending belt of the Continental Divide, and the mountain-park province of South Park. Altitudes range from 9,000 to more than 14,000 feet.

The principal divisions of the bedrock are: (1) Pre-Cambrian rocks, including whitish-gray and pink granites, injection gneiss, schist, and pegmatite; (2) pre-Pennsylvanian dolomite, quartzite, conglomerate, and shale; (3) Pennsylvanian and Permian sedimentary strata of diverse lithologic types, the lower third of which is prevailingly nonred and the upper two-thirds prevailingly deep red, though a local small patch of the upper strata have turned green by contact metamorphism; (4) the Morrison formation and Dakota sandstone (Upper Jurassic and Upper Cretaceous); (5) the Benton shale and the Montana group (Upper Cretaceous); (6) the Laramie formation; (7) the Tertiary strata; (8) early Tertiary prophyries of the mountain province; and (9) Tertiary igneous rocks of the Mountain Park province of South Park. Of these rocks, the pre-Cambrian rocks, the Tertiary igneous rocks, and the sandy facies of the sedimentary strata are the most resistant to destruction by agencies of erosion. Pebbles of the pre-Cambrian rocks and the green contact-metamorphosed rocks have proved especially useful guides in tracing the sources of unconsolidated deposits containing gold. The principal sources of placer gold have been in a mineralized area near the head of Middle Tarryall Creek and in a northeasterly mineralized belt crossing the Mosquito Range.

Physiographic interest in the mountain province is focused particularly on the valleys. Valley erosion during a long preglacial period established the general drainage pattern, and subsequently ice action greatly modified the valley form and controlled deposition of unconsolidated materials in all the principal valleys. Postglacial erosion has produced only insignificant changes. The last glaciers, of Wisconsin age, deposited moraines that now are cut through by narrow stream gorges but are otherwise almost completely uneroded. From these moraines the positions not only of maximum advance but also of important halts or readvances of the glaciers can be deduced. Evidence of two major and numerous minor ice stands is found in each large valley. In this report the times of maximum ice advance and of the two principal ice stands during retreat are called the Fairplay, Briscoe, and Alma substages, respectively, of the Wisconsin stage of glaciation. The older moraines, however, have been swept away by the later ice movements everywhere except above the sides of the glaciers of Wisconsin age and downstream from their ends. Even where not removed by ice, the older moraines have commonly been partly or wholly removed by normal erosion.

At the time of maximum advance, the last glacier in the South Platte Valley terminated half a mile upstream from Fairplay. It was joined by tributaries from Buckskin, Mosquito, and Sacramento Gulches, as well as smaller gulches, and near its lower end it spilled across its eastern divide into Beaver Creek. Moraines marking the two principal later stands of the glacier were deposited near the Briscoe ranch and Alma, respectively. The terminus of the Middle Tarryall Glacier at its farthest advance during the Wisconsin stage of glaciation was at the Tarryall confluence. It was joined by a major tributary from Montgomery Gulch, and its lower portion spilled over into North Tarryall Creek. Two principal stands during retreat have been correlated with those of the South Platte Valley. Fourmile Glacier and several small glaciers having little or no economic significance are briefly discussed.

In South Park the remnant of a gravel-covered, planate valley floor, much older than the earliest glacial stage, forms part of the divide between the South Platte and Fourmile Valleys. It is mapped as terrace No. 1. Each of these valleys contains remnants of a considerably younger, though preglacial, gravel-covered erosion surface, mapped as terrace No. 2. Both terraces represent temporary base levels of erosion. Alternating erosion and deposition until the close of the glacial period has resulted in great dissection of terrace No. 2 and in deposition of outwash aprons extending downstream from the Wisconsin moraines. The apron in the South Platte drainage area constitutes most of the present valley floor. It has been slightly dissected by postglacial erosion, with development of two benches in addition to the present stream channel. The Fourmile outwash apron lies entirely south of the present stream. All deposits from both the South Platte and Fourmile Valleys contain an abundance of pre-Cambrian boulders. Materials derived from Beaver and Crooked Creeks, on the contrary, contain no pre-Cambrian rocks.

Tarryall, Park, Michigan, and Jefferson Creeks within South Park constitute a single large drainage basin. Terraces corresponding to the two preglacial terraces of the South Platte have been found and mapped. The Tarryall terrace No. 2 is preserved over a large area. One stage in dissection of Terrace No. 2 is preserved by a moderately extensive terrace mapped as terrace No. 3 and correlated with the pre-Wisconsin interglacial stage. Outwash aprons extend, as in the South Platte, downstream from Wisconsin moraines. Materials from the head of Tarryall Creek are characterized by an abundance of green contact-metamorphosed rock and an absence of pre-Cambrian rock; those from Michigan and Jefferson Creek, conversely. Materials from Trout Creek contain only minor amounts of green contact-metamorphosed rock.

Placers have yielded over 4 million dollars in gold, of which three-fifths came from South Platte Valley, somewhat less than two-fifths from Tarryall Creek, and virtually all the remainder from Beaver Creek. The Tarryall gold exceeds 0.858 in fineness, is associated with green contact-metamorphosed rock, and was derived from the Montgomery-Deadwood mineralized area. The South Platte and lower Beaver Creek gold has a fineness of less than 0.837, is associated with pre-Cambrian rock, and was derived from the mineralized belt that crosses the Mosquito Range.

By far the most productive materials have been the outwash aprons extending away from Wisconsin moraines in all these valleys. Next in productivity are the outer parts of the moraines themselves; the lowermost moraines are far more productive than any formed during retreat, but those of the Alma, substage of the Wisconsin rank next. Lateral moraines have been productive at some places.

The last part of the report is devoted to a detailed discussion of individual placers or groups of placers. Known deposits are described and chances for further discoveries evaluated for each locality. In 1939 the largest undeveloped areas to warrant prospecting were the outwash aprons of South Platte and Tarryall Valleys, but in addition much of the remainder of Tarryall Creek outside the mountains offers some promise of future production. During 1939–42 the outwash aprons were extensively prospected and partly exploited.

INTRODUCTION

SCOPE OF THE WORK

Gold placers not only attracted the earliest settlers but also have provided a nearly continuous, though at times very small, source of revenue in northwestern Park County. Since 1934 renewed interest has given them a prominent place in mining activity. Almost every conceivable method of working has been applied at one place or another within the district.

During the period 1928–38, information conerning placers was obtained incidentally to work on the bedrock geology and lode deposits of the Mosquito Range and the Beaver-Tarryall district. This preliminary information led to a belief that further, intensive study might reveal undeveloped resources of gold-bearing gravels. Accordingly, nearly 3 months of field work was done during the summer of 1939 as a part of the Geological Survey's cooperative program with the State of Colorado. This field work revealed large areas of promising outwash gravel in South Platte and Tarryall Valleys and of moraine in South Platte Valley. After 1939 private individuals and companies extended their prospecting into these areas, and several new placers were started, but this later work is not included in the present report. The total gross output during the period 1939–42 amounted to about $2,000,000.

The bedrock geology is a starting point for consideration of the placers. The character, durability, and distribution of different types of rock are pertinent to an understanding of the reasons for past and present topographic forms and to a knowledge of the sources of materials making up the unconsolidated deposits. The locations of lode deposits that contributed the gold are equally relevant to the problem. The section dealing with bedrock geology in this report makes free use of previously published work of the writer,[1] of a report by Washburne,[2] and of a then unpublished map of South Park by a group of geologists from Northwestern University.

[1] Singewald, Q. D., Stratigraphy, structure, and mineralization in the Beaver-Tarryall area, Park County, Colo.; a reconnaissance report: U. S. Geol. Survey Bull., 928–A, 44 pp., 1942. Singewald, Q. D., and Butler, B. S., Ore deposits in the vicinity of the London fault of Colorado: U. S. Geol. Survey Bull. 911, 74 pp., 1941.

[2] Washburne, C. W., The South Park coal field, Colorado: U. S. Geol. Survey Bull. 381, pp. 307–316, 1910.

An understanding of gold distribution depends upon a thorough knowledge of the physiography which involves not only the present topography but also that of past geologic periods as well. Within the mountain province, physiographic interest is focused primarily upon the valleys, but in South Park the entire area must be considered; furthermore, glaciers occupied all the major valleys of the mountain province but ended approximately at the topographic boundary between the mountains and South Park. For these reasons it has been convenient to describe the two provinces separately in this report. Accordingly, the writer first deals briefly with the bedrock geology, then discusses the physiography of each province in detail, and finally describes the gold placers and their relations to other geologic features. Suggestions for prospecting possible gold-bearing ground are given for each of the main valleys and their large tributaries.

In the treatment of glacial geology, free use has been made of the work of Capps [3] in the Leadville quadrangle.

GEOGRAPHY

The area consists of more than 250 square miles in the northwest corner of Park County. Extending southward for 17 miles from the Continental Divide, the area is bordered on the west by the crest of the Mosquito Range and on the east approximately by Jefferson Creek. Latitude and longitude are shown on the geologic maps.

Fairplay, the principal town, is 85 miles by highway from Denver, 62 miles from Colorado Springs, and 39 miles from Buena Vista. Other towns are Alma, Como, and Jefferson. Aggregate population of the entire area is less than two thousand.

There are two principal drainage systems, which eventually unite far to the southeast of the area mapped. The South Platte River, which flows south-southeastward past Alma and Fairplay, heads against the Continental Divide. It is joined by Buckskin, Mosquito, and Sacramento Creeks, and several smaller creeks from the west and by Beaver and Crooked Creeks from the east. Two other creeks, Fourmile to the west and Trout to the east, join the South Platte River a few miles beyond the area mapped.

The other drainage system, mainly in South Park, comprises Jefferson, Michigan, and Tarryall Creeks, all of which head at the Continental Divide and have numerous tributaries. Jefferson joins Michigan Creek near the east border of the area mapped; Michigan in turn joins Tarryall Creek a few miles farther southeast.

The topography comprises a mountain province and a mountain-park province. The mountain province may be divided into two parts,

[3] Capps, S. R., Jr., Pleistocene geology of the Leadville quadrangle, Colorado: U. S. Geol. Survey Bull. 386, 99 pp., 1909.

the northward-trending Mosquito Range, and the northeastward-trending range along which the Continental Divide crosses from the Park Ranges to the Front Range. The two ranges join at Hoosier Pass and are divided southward from that point by the South Platte Valley. The mountain-park province can best be regarded as subdivided by relatively low ridges that form divides between principal streams. The main divides are those separating Fourmile, South Platte, Trout, and Tarryall Creeks. In addition, there is a noteworthy range of hills cut through by Michigan and Tarryall Creeks in the eastern part of the region.

Altitudes range from a little more than 9,000 feet to more than 14,000 feet. Several of the highest peaks are neither on the Continental Divide nor on the crest of the Mosquito Range. Climate is comparable to that of other regions of Colorado at the same altitude. Below timber line, at approximately 11,500 feet, pine and aspen are abundant in the mountains but scarce in South Park.

EXPLANATION OF MAPS

Geologic features of interest are too numerous and overlapping to be included on a single map. Plate 8, depicting the bedrock geology, shows only features pertinent to discussion of the placers and is not to be regarded as a substitute for more detailed maps published elsewhere. Boundaries in the mountain province are generalizations from work of the writer, whereas those in South Park are taken, with modifications, from the work of Washburne. It should be emphasized that the writer made no study of bedrock in South Park except incidentally to his physiographic work.

Plate 9 has been compiled for the convenience of the reader, particularly in reference to the section on glacial geology. Boundaries of ice during the last glacial stage in the South Platte Valley and all its western tributaries including Fourmile Gulch conform closely with those depicted on Capps' map, but some additional detail has been inserted. Ice boundaries mapped in the other valleys are based entirely on the writer's interpretation.

Plate 10 contains the information most directly related to the placers. All active or abandoned workings that existed in 1939, as well as their physiographic settings, are shown. Most of the work was recorded directly on topographic base maps of the United States Geological Survey, but a small area mapped by car traverse south of Fairplay has been appended.

In addition to the general maps, three large-scale maps of local areas are included in the report. Each shows locations and depths to bedrock in prospect holes and pits; this information was furnished by individual operators. Surface elevations recorded on plate 11 are based on aneroid readings made by Mr. Frederick Gros, whereas those on plates 12 and 13 have been obtained in a plane-table survey by the writer.

The index map (fig. 5) shows general location of the area.

ACKNOWLEDGMENTS

The writer is particularly indebted to Mr. B. S. Butler, who supervised the work, participated in collecting all the preliminary information, and spent 8 days of 1939 in the field discussing most of the problems dealt with in this report. The writer likewise is greatly

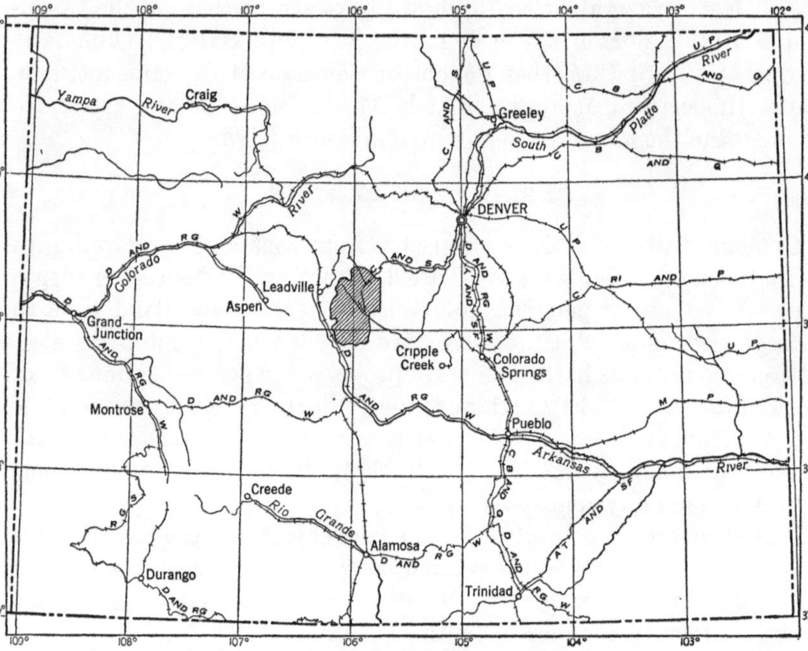

FIGURE 5.—Map of Colorado showing location of area described in this report.

indebted to Mr. S. R. Capps, not only for the value of his publication but also for advice given in the field during 2 days of 1940. Mr. Charles W. Henderson, of the United States Bureau of Mines furnished invaluable aid in many ways. Local operators and prospectors have been most generous in providing information; all cannot be named, yet special thanks are due Messrs. Edward S. Ames, E. J. Cunningham, R. W. Derby, Jr., and John Singleton. The writer also is grateful to Messrs. W. S. Burbank and G. F. Loughlin for valuable suggestions regarding the manuscript. The able assistance of Mr. Frederick Gros is gratefully acknowledged. The writer wishes to express great appreciation of the generosity of Mr. Charles H. Behre, Jr., and all the other members of the Northwestern University group in providing a small-scale photostat of their then unpublished map of South Park.

BEDROCK GEOLOGY

METAMORPHIC, SEDIMENTARY, AND IGNEOUS ROCKS

PRE-CAMBRIAN IGNEOUS AND METAMORPHIC ROCKS

The pre-Cambrian rocks include granite, injection gneiss, schist, and pegmatite. All are relatively resistant to erosion and to destruction during transportation. The granites, the pegmatites, and the more highly granitized facies of injection gneiss are particularly durable.

The granite is of two types. The one is prevailingly whitish-gray and the other pink, owing to colors of the characteristic feldspars. The former has been correlated with the Silver Plume granite, and the latter, which is the coarser grained, with the Pikes Peak granite. The injection gneiss and the schist are closely related. They form a transitional series of schistose rocks containing from minor amounts to an abundance of injected granite and pegmatite. The pronounced foliation and decidely darker color readily serve to distinguish these rocks from granite. The pegmatite is very coarse-grained and may be either white or pink; it is much less abundant than any of the other rocks.

Whitish-gray granite and the injection gneiss series, accompanied by much pegmatite but very little pink granite, crop out over extensive areas in the upper parts of South Platte, Buckskin, Mosquito, and Fourmile Valleys, and in smaller areas of Lincoln, Moose, and Sacramento Gulches. (See pl. 8.) Consequently, these rocks are conspicuous and diagnostic constituents of all unconsolidated deposits derived from these localities. The same rock types are abundant in the unmapped, upper parts of Michigan and Jefferson Creeks, and are therefore important constituents in unconsolidated deposits found along these creeks also.

Pink granite with pegmatite, but unaccompanied by schist, injection gneiss, or white granite, crops out in a range of low hills in the extreme eastern part of the area and also in several outliers to the northwest. Unconsolidated deposits derived from them, therefore, are rich in granite and pegmatite yet devoid of schist or injection gneiss.

Pre-Cambrian rocks do not crop out at the head of Tarryall, Trout, Crooked, or Beaver Creek, and are entirely absent from the unconsolidated deposits along these creeks. Pre-Cambrian rocks are also absent along the southern divide of Sacramento Creek and the northeastern divide of Fourmile Creek.

PRE-PENNSYLVANIAN SEDIMENTARY ROCKS

Pre-Pennsylvanian sedimentary rocks constitute a series—ranging from 200 to 600 feet thick—of dolomite, some quartzite, and small amounts of conglomerate and shale. They are on the whole more resistant to erosion than any other group and so tend to form prominent cliffs. White fine-grained Cambrian quartzites are especially resistant to destruction, and the dolomites are moderately resistant.

These pre-Pennsylvanian rocks crop out in two main belts (see pl. 8), one on either side of the London fault. Their quartzites and dolomites, therefore, are represented in deposits of South Platte, Buckskin, Mosquito, Sacramento, and Fourmile Gulches but are less abundant than pre-Cambrian rocks because their exposed area is much smaller.

Bedrock has not been mapped in the upper parts of Michigan and Jefferson Creeks, so it is not known whether pre-Pennsylvanian sedimentary rocks are present there or not. None were noticed, however, in deposits along these creeks.

PENNSYLVANIAN AND PERMIAN SEDIMENTARY ROCKS

General character.—Overlying the pre-Pennsylvanian rocks and cropping out to the east of them is a series of Pennsylvanian and Permian sedimentary rocks whose maximum thickness amounts to about 10,000 feet. Their component strata are nearly all clastic, consisting of shale, siltstone, sandstone, and conglomerate, but contain many thin intercalated beds of limestone and dolomite. All the clastic strata except a few of the basal beds are conspicuously micaceous and contain an abundance of feldspar. Some conglomerates contain cobbles up to 6 inches in diameter. Almost throughout, the beds are lenticular and abruptly alternate across the bedding from coarse- to fine-grained facies.

Color, although of little use in establishing a stratigraphic boundary, provides the most distinctive lithologic division of the series, for approximately the lower third is nearly free of red beds, whereas most of the upper two-thirds is deep red. The boundary between the prevailingly red and the prevailingly nonred parts is concealed over most of the area, yet its approximate position is sufficiently useful in tracing the sources of unconsolidated materials to warrant representation on plate 8.

Except where bolstered by an abundance of intrusive rock, the series is rapidly eroded and so tends to crop out in the lower areas. Some of the more sandy members, nevertheless, individually resist destruction fairly well and so are represented in unconsolidated deposits. A few thin quartzite beds near the base closely resemble the Cambrian quartzites.

Prevailingly nonred part.—The prevailingly nonred part of the series crops out in a continuous belt, as shown on plate 8, from the Continental Divide to the south margin of the map, and also in a somewhat narrower belt east of the London fault. Its materials, therefore, are fairly abundant in deposits south of Mosquito Creek and in Beaver Valley; they are minor constituents of deposits in the lower parts of South Platte, Buckskin, and Mosquito Creeks.

Red beds.—The prevailingly red part of the series, commonly called red beds, crops out over a wide area extending from the Continental Divide to the southern border of the area mapped. Though poorly resistant to destruction, the large quantity of them available makes these rocks, particularly the more sandy facies, conspicuous constituents of deposits along streams that drain them.

Green contact-metamorphosed rock.—Of particular value as an aid in tracing the sources of gravel and gold are the green rocks derived from an area of contact metamorphism surrounding a small stock at the heads of Montgomery and Deadwood Gulches. They crop out within an area that closely coincides with the Montgomery-Deadwood area of lode mineralization shown on plate 8, and also in a small outlying area west of Boreas Pass. The green color is due to developments of chlorite and epidote, accompanied by elimination of red ironstain during metamorphism. Although the same color due to the same minerals is widespread in other rocks of the region, the rocks from the Montgomery-Deadwood area can readily be recognized. They have a thinly laminated structure and a sedimentary texture, in contrast to the igneous or schistose textures of all other green rocks; moreover, the finer-grained facies from the margins of the area show interlaminated green and red colors.

For convenience in later descriptions the rocks from the Montgomery-Deadwood area will be called "green contact-metamorphosed." Being much more durable than the red beds from which they were derived, they are fairly abundant constitutents of all deposits derived from Montgomery, Deadwood, and Middle Tarryall Gulches. They are minor constituents, moreover, of materials from the heads of Beaver, Crooked, and Trout Creeks and from North Tarryall Creek above its confluence with the Middle Tarryall. They are absent from deposits derived from Jefferson, Michigan, South Platte, Buckskin, Mosquito, Sacramento, and Fourmile Gulches.

MORRISON FORMATION AND DAKOTA SANDSTONE (UPPER JURASSIC AND UPPER CRETACEOUS)

A basal sandstone of the Morrison formation is overlain by limestones and variegated shales of that formation which in turn are overlain by Dakota sandstone. Total thickness of the two formations is about 650 feet, of which sandstone constitutes nearly half. Both sandstones are quartzose, fine-grained, and white except where stained during weathering.

The sandstones of the Morrison and Dakota formations are far more resistant than the sedimentary beds immediately above or below. Consequently, in South Park they form an unbroken escarpment southward from Trout Creek. Northward from Trout Creek they rise to the crest of Little Baldy Mountain, from which a western belt continues without an escarpment to the Continental Divide, but a much shorter eastern fork is cut off against a large stock east of Tarryall Creek.

The Dakota sandstone and to a less extent the sandstones of the Morrison formation resist destruction and are therefore conspicuous in deposits downstream from outcrops of these formations along Tarryall and Trout Creeks. They are also conspicuous in deposits along Michigan Creek, which heads near Mount Guyot, north of the area mapped. As applied to constituents of gravel, the term "Dakota sandstone" will for convenience be used to imply a mixture of both Dakota and Morrison rocks but with Dakota predominating.

BENTON SHALE AND THE MONTANA GROUP (UPPER CRETACEOUS)

The Benton shale, which overlies the Dakota sandstone, is a dark, rather fissile shale that is very easily eroded. It does not survive transportation and therefore is absent or rare in unconsolidated deposits. The small quantities that are preserved come mostly from the vicinity of large intrusives, where the shale has been baked.

Neither the stratigraphic nor the lithologic details of the Montana group are pertinent to this study. The rocks constituting it are mainly yellow-weathering siltstones accompanied by shales and sandstones. All are easily weathered and so friable that they do not survive transportation. The Montana group, together with the Benton shale, crops out in a large arc east of the escarpment of the Dakota sandstone.

LARAMIE(?) FORMATION (UPPER CRETACEOUS?)

According to Washburne,[4]

> the coal beds of South Park occur in what is presumably the "Laramie" formation, consisting of sandstone with subordinate beds of carbonaceous shale and ranging in thickness from 375 feet down to the vanishing point . . . The formation . . . rests on yellow sandstone containing upper Montana marine fossils and is unconformably overlain by conglomerate beds that are undoubtedly equivalent to part, at least, of the Shoshone group.

The sandstones of the Laramie(?) formation, together with the overlying conglomerate, resist erosion somewhat more effectively than

[4] Washburne, C. W., The South Park coal field, Colorado: U. S. Geol. Survey Bull. 381, p. 308, 1910.

beds for a considerable stratigraphic distance above and below, and so tend to crop out as a low ridge. This ridge is particularly prominent, however, only where bolstered by igneous rocks. Pebbles derived from the Laramie(?) formation are not constituents of gravels.

ARKOSE OF TERTIARY AGE

Sedimentary rocks of Tertiary age crop out over a large area (see pl. 8) southeast of the main belt of the Laramie(?) formation. Above a thin zone of coarse conglomerate with many porphyry boulders, the strata are predominantly arkose composed of quartz, rather fresh feldspar, and pegmatite. Most grains are less than a quarter of an inch in diameter, but some are as much as half an inch. A few layers contain friable pebbles, cobbles, or boulders up to 1 foot in diameter. Intercalated at least in the lower part are lenticular beds of green, highly micaceous, and slightly pebbly shale ranging from 1 foot to 10 feet in thickness. Along Tarryall Creek these beds are conformably folded with the Larmie(?) and so must be older than the regional, Laramide structures.

The rocks above the basal conglomerate are very friable and hence are eroded as readily as those beneath the Laramie(?) formation. None of the materials from them could be transported very far without destruction, yet they are abundant in certain gravels formed almost in place within their outcrop area.

TERTIARY IGNEOUS ROCKS

Porphyry.—The term "porphyry" is used here for convenience in referring to any or all representatives of the White porphyry and the Gray porphyry group so familiar to miners in the Leadville district. The White porphyry may be devoid of distinguishable minerals but more commonly has a few grains of quartz, mica, and feldspar enclosed in a dense-textured, whitish-gray groundmass; it weathers white or pale rusty brown. The Gray porphyry group, which includes several different varieties, normally has an abundance of visible minerals enclosed in an inconspicuous, gray groundmass. The identifiable minerals include light gray feldspar, black biotite and hornblende, or their alteration products, and quartz. Some varieties are quartz-poor, other quartz-rich; and one variety contains large pink orthoclase crystals. Locally, alteration has developed so much chlorite and epidote that the rock is green; at a few places the rock has been bleached. Upon weathering, the rock becomes light gray but stained more or less by rusty-brown iron hydroxides. Where it occurs in stocks, the Gray porphyry group may be devoid of matrix.

To show the outcrops of White porphyry and the Gray porphyry group would unduly complicate plate 8. They are abundant as sills and dikes throughout the mountains, yet they are almost entirely absent in South Park. They are more resistant to erosion than any of the sedimentary rocks except quartzite and hard sandstone, and about equal in resistance to the pre-Cambrian rocks. Owing to their wide distribution and durability, the porphyries are conspicuous constitutents of all unconsolidated deposits and therefore are of no diagnostic value in tracing materials to their source.

Three stocks composed of igneous rocks closely related to the porphyries are shown on plate 8.

Other types.—Two petrographic types differing markedly from the porphyries of the mountain province occur in South Park. The one is a fine-grained, tuffaceous-looking whitish-gray rock that is not represented on plate 8 but is interlaminated with strata of post-Laramie age and perhaps also with strata of the Laramie(?) formation. It resists erosion slightly more than the sedimentary rocks. The other type is dark, brownish gray. It resists erosion sufficiently to form a few moderately high hills in T. 9 S., R. 76 W. Other small exposures are not shown on plate 8.

Both the light- and the dark-colored types occur entirely outside areas in which the writer has mapped bedrock geology, and they have no bearing on gold distribution. Their character, age, and relations, therefore, have not been studied in connection with the present work. They have been mapped by the Northwestern University group of geologists.

STRUCTURE

The area under consideration lies on the east flank of the northward-trending Sawatch arch,[5] so the prevailing dip of the formations exposes successively younger beds eastward. This simple picture, however, is complicated by six major structural features trending more or less parallel to the arch; five are shown on the geologic map (pl. 8). In the Mosquito Range, east of its crest, three major faults, all with upthrow on the east, have ruptured the steep west limbs of long and narrow anticlines causing a repetition of part or all of the rock sequence. Extending from Hoosier Pass southward at least to Alma a zone of reverse dips and faults indicates a major structural feature but does not cause repetition of a major rock series; it is not represented on plate 8. A fault extending from Little Baldy Mountain northward to Boreas Pass causes repetition of red beds on its east side. All these faults have been mapped by the writer previously.

In South Park, the anomalous position of beds of the Laramie(?) formation north of Como led Washburne [6] to postulate the South Park

[5] Butler, B. S., Relation of the ore deposits of the southern Rocky Mountain region to the Colorado Plateau: Colorado Sci. Soc. Proc. vol. 12, p. 28, 1929.

[6] Washburne, C. W., op. cit., pl. 16, p. 314.

fault, as shown on plate 8. A much larger fault, originally mapped by Washburne and more recently by the Northwestern University group of geologists, separates granite hills along the east border of the area from sedimentary beds to the west. The Northwestern University geologists regard this fault as a low-angle thrust and plan to describe several isolated outliers,[7] or klippen, of the overthrust granite body that are found in the Hartsel region, 15 to 20 miles southeast of Fairplay. The writer believes that a rounded ridge northwest of the Fairplay-Denver highway, between Jefferson and Michigan Creeks, is another granite outlier separated from underlying arkose of post-Laramie age by a thrust fault. The arkose is exposed in highway cuts at the south tip of the ridge.

No attempt was made by the writer to decipher the remainder of the South Park structure.

LODE DEPOSITS

All the lode deposits from which placer gold could be derived are located within the mountains, and nearly all are concentrated within definite areas or belts of mineralization.

The only mineralized area east of South Platte Valley lies at the heads of Montgomery and Deadwood Gulches. Its approximate outline, which coincides with that of the green contact-metamorphosed rocks, is shown on plate 8. The deposits, which are of a high-temperature (pyrometasomatic) type, surround the apex of a small stock. In the absence of master channels, the ore-forming solutions spread into innumerable small fissures to deposit lodes containing small quantities of gold and almost no copper, lead, zinc, or silver. Though individually not favorable for commercial exploitation, these lodes in aggregate have supplied large quantities of placer gold.

In the Mosquito Range the greatest ore deposition, as well as the greatest igneous intrusion, took place within a belt that is approximately 3 miles wide, is slightly convex toward the southeast, and trends northeastward from Leadville to North Star Mountain. Scattered lodes occur throughout the belt, but the vast majority are concentrated within four mineralized areas.

The North Star Mountain area (see pl. 8), which is the only one not related to northward- or northwestward-trending structures, has supplied chiefly gold and minor amounts of silver and other metals. In this area there are many veins in both the pre-Cambrian metamorphic and the overlying pre-Pennsylvanian sedimentary rocks, but most of them are too small to be mined; nevertheless, the distribution of placers in South Platte Valley indicates that the North Star mountain

[7] Stark, J. T., oral communication.

area and perhaps also scattered lodes to the west were a major contributor of placer gold.

A fairly large mineralized area surrounding Mount Bross and Mount Lincoln includes three major silver mines, as well as many small mines and prospects that have silver and only insignificant amounts of gold. What little gold has been found, occurs chiefly in the northern part of the area which doubtless contributed a minor amount of gold to placers.

An area extending from the north side of Buckskin Gulch across lower Loveland Mountain to the south side of Mosquito Gulch includes several medium-sized gold mines as well as many small mines and prospects. The output has consisted mainly of gold accompanied by minor amounts of silver and other metals. The lodes are very similar to those of North Star Mountain, so they may possibly have been a source of placer gold. That some gold was derived from this source is proved by small placers in lower Buckskin Gulch. Perhaps the fairly numerous lodes farther up Buckskin Gulch, near the small stock shown on plate 8, also contributed.

The fourth mineralized area of the belt is located in the vicinity of London Mountain. It includes the London group of mines, greatest commercial producers of the entire district, whose output has consisted chiefly of gold, with minor amounts of silver, zinc, lead, and copper. Because of the very large yield from lode mines, this area might at first be regarded as a major contributor of placer gold; nevertheless, no appreciable quantities of placer gold from it have yet been discovered. Throughout most of this area the output of lode ore has come from veins or replacement bodies that either do not crop out at all or else have been only slightly eroded. It is quite possible, however, that veins formerly existed within the eroded area of South Mosquito Gulch, and the London Mountain area must therefore be regarded as a possible though not certain source of placer gold.

All the lodes within the mineralized belt of the Mosquito Range east of its crest were deposited at intermediate temperatures, well below those in the Montgomery-Deadwood area. Small quantities of magnetite at North Star Mountain suggest that the temperature of ore deposition here was a little higher than at other areas.

Two areas of mineral deposits south of the main belt of mineralization are associated with major faults, shown on plate 8. The deposits were formed at temperatures lower than those that prevailed farther north and have supplied silver ore with little or no gold. They could not, therefore, have contributed noteworthy quantities of placer gold.

PHYSIOGRAPHY OF THE MOUNTAIN PROVINCE

SEQUENCE OF EVENTS

The physiographic history of the mountain province may conveniently be divided into three periods: Preglacial, glacial, and postglacial. The preglacial period was one of normal erosion following the Laramide revolution. During that time the drainage pattern was evolved and the broader topographic features were established. Then followed a glacial period during which ice action greatly modified the form of all the larger valleys. The topography of the upland areas between glaciers, on the contrary, merely continued in its evolution, much as before. The short postglacial period has witnessed relatively little change.

Preglacial weathering, erosion, transportation, and deposition within the mountains has had only an indirect bearing on the placer problems. During preglacial time the thick covering of rocks that overlay the gold lodes was removed, then the lodes themselves were attacked. All the preglacial valleys in which detrital gold accumulated, however, have been glaciated, with consequent redistribution of the gold. Present placers within the mountains therefore bear close relationship to the glacial deposits. For these reasons the unglaciated areas, as well as preglacial events in the glaciated areas, will not be considered further.

GENERAL RESULTS OF GLACIATION

GLACIAL EROSION AND DEPOSITION

Cause.—Climatic changes over all of North America initiated glaciation in favorable places. As the snow line of the mountains reached lower altitudes, increasingly large amounts of perennial snow accumulated in depressions and valley heads, became converted into ice, and began to move slowly downhill. Contributions from many ice fields converged toward a valley to form a thick ice sheet, called a glacier, which in turn flowed slowly down the valley beyond the snow line before melting. So long as the snow line reached lower and lower altitudes, the ice front in each valley advanced, but when the snow line retreated to higher levels, the ice front also retreated.

Erosion.—By a combination of plucking and abrasion, glaciers may erode much more rapidly than streams. At their heads they tend to develop cirques—semicircular hollows having precipitous walls and relatively smooth floors. Enlargement of two nearby cirques may reduce the intervening divide to an extremely sharp, serrated ridge. Below the cirques, ice erosion widens and deepens the valleys, forming U-shaped profiles in cross section. The actual amount of deepening depends on many factors; in a general way, the more irregular the original surface, the more highly fractured the bedrock, and the thicker the ice, the deeper will be the excavation. For these reasons, main valleys were commonly deepened more than their tributaries. Thus, glacial ice tended not only to sweep away all unconsolidated material in its path but cut through the weathered, "rotten" bedrock for a considerable depth into resistant, unweathered rock.

Deposition.—The great quantity of materials eroded by a glacier is deposited either in the form of moraines directly by the glacier or as outwash aprons by streams issuing from it. Moraines have distinctive form and character. Generally they have an extremely irregular, hummocky surface and include numerous depressions having no outlets. Their materials, unsorted as to size, are a mixture of all the rocks over which the glacier has passed; the less durable rocks tend to be ground to a fine powder, the more durable to remain as boulders. Gold, if present in the source material will be included in this debris.

A considerable part of the debris is deposited at the lower end of the ice. It is continuously pushed forward and in part overridden while the ice advances but is left behind during retreat of the ice. Consequently, the position of maximum advance is generally clearly demarked by a crescent- or horseshoe-shaped terminal moraine. Where best developed, it forms a sharp-crested, but hummocky ridge; the outer or downstream side rises abruptly from an outwash apron, whereas the inner side is bordered by a morainal bench. Where two stages of approximately equal advance were interrupted by a brief recessional stage, double-crested moraines were formed. Such moraines, though well-defined elsewhere in Colorado, were not formed in this region. Where poorly developed, the terminal moraine may consist of only a few low hummocks extending across the valley. Under any circumstances, however, the terminal moraine opens upstream into a broad, relatively flat basin which may have been occupied by a lake of glacial water impounded by the moraine. In such places overflow over the morainal dam may start stream cutting, which, proceeding rapidly in the unconsolidated debris, eventually cuts a channel which completely drains the lake. Naturally, such interior basins vary in detail, their shape being influenced by the height of the moraine and the width of the valley.

Secondary terminal moraines, located at intervals upstream from the point of maximum advance, occupy all except the smallest glaciated valleys shown on plate 9. Though differing in size and detail of form, each secondary terminal moraine is characterized by an arc that crosses and clogs the valley, and each opens upstream into a well-defined "interior" (former lake) basin. Each represents a more or less prolonged stand of the ice front, produced either by a halt in retreat or, more probably, by a readvance of appreciable magnitude. Elsewhere, valley floors are covered by debris dropped by retreating ice.

Lateral moraines, as their name implies, are deposited along the sides of a glacier and merge downstream with the terminal moraine. Lateral moraines deposited at the time of maximum advance commonly form benches against the rock walls of the valleys. At places, however, where ice spilled over one or both divides along the lower reaches of adjoining valleys, the lateral moraine forms a sharp-crested hummocky ridge overlying the original bedrock divide and slopes off in both directions, on the outer side as far as the former ice margin. Secondary lateral moraines, formed during prolonged ice stands or readvances subsequent to maximum ice advance, are clearly present in places, but their record is in general less complete than that of the secondary terminal moraines. Only the secondary lateral moraines that can be identified with considerable assurance have been indicated on plate 9; nearly all of them are attached to well-defined secondary terminal moraines. These lateral moraines form benches, some against slightly older moraine, others against bedrock, but along all of them moraine deposited earlier in the same glacial stage can be found on the hillside above the bench.

Where glaciers of two adjacent valleys coalesced, the junction of the two ice lobes generally is indicated by a line of debris, called a medial moraine, which is a continuation of the adjoining lateral moraines.

Ice-border channels.—Ice-border channels, though not formed by ice erosion or deposition, are closely associated with moraines. Huge quantities of water issuing from the sides and front of melting ice may form a wide continuous sheet or be confined to definite channels, which are generally braided. Being overloaded with debris from the glacier, these streams deposit their materials as an outwash apron. The stream gradients over short distances immediately adjacent to the ice, however, may be steep enough to cause a slight amount of erosion, or at least to prevent deposition in the channel between the ice on one side and the valley wall on the other. After retreat, the wall of ice is replaced by one of moraine, so that the abandoned, dry channel remains.

Ice-border channels are especially well developed at approximately the point where a terminal moraine merges with a lateral moraine. They may be recognized along the outer border of nearly all the moraines shown on plate 9, though the 50-foot contour interval is too great to picture them clearly. At a few places they constitute more tangible evidence of an ice stand than does the terminal moraine itself.

STAGES OF GLACIATION

Glaciation recurred in several stages and substages throughout North America, but in the region under consideration only that of the latest or Wisconsin stage is clearly recorded. The deposits of this stage are therefore described at some length; those of earlier glacial and interglacial stages are described to the extent that fragmentary information on them permits.

WISCONSIN STAGE

Moraines.—Throughout the mountainous part of the region a series of well-preserved moraines is found in every valley that contained a glacier of moderate or large size. Each moraine has been cut by a relatively narrow gorge occupied by the present stream but otherwise has undergone no appreciable change in form; hence, all such moraines must be fairly recent and belong to a single stage of glaciation. This stage, which is well-defined throughout the glaciated part of North America, has been named the Wisconsin stage.

The maximum advance of ice during the Wisconsin stage is indicated in each valley by the lowermost well-preserved moraine. Farther upstream secondary terminal moraines, or ice-border channels devoid of associated terminal moraines, indicate noteworthy ice stands during retreat. In the largest valley, that of the South Platte, there are two unusually large secondary moraines, which contain placer gold. They are here named the Briscoe moraine and the Alma moraine. The time covered by the deposition of these secondary moraines is to be known as the Briscoe substage and Alma substage, respectively. In Tarryall Valley likewise there are two particularly conspicuous secondary moraines with which gold placers are associated; presumably they may be correlated with the two in the South Platte. The probable equivalents of these moraines in other valleys have been indicated on plate 9.

From their form and character alone it would not be possible to state whether the Briscoe and Alma moraines represent merely long ice stands, slight readvances, or renewed glaciation following complete retreat. The presence of considerable placer gold within the Alma moraine and its equivalent in Middle Tarryall Gulch, however, suggests that a considerable period of weathering immediately preceded its deposition, and that it is therefore a product of renewed glaciation rather than of a mere halt in ice retreat. This interpretation depends on an assumption that gold freed by weathering prior to the Wisconsin stage had been swept away and deposited in and beyond the lowermost terminal moraine at the time of maximum advance. Evidence from distribution of the gold, which will be discussed in detail in a later section of the report, is less conclusive with regard to the

Briscoe moraine, but this moraine probably also represents a readvance. These two readvances, therefore, may appropriately be referred to as substages. Perhaps it should be stressed, however, that any time interval between the substages, as proved by the preservation of moraines, was of very short duration compared with the time separating the entire Wisconsin stage from the preceding stage.

Snow lines.—Glacial geologists working in the Alps apparently have achieved some success on calculating the position of ancient snow lines on the basis of two relations observed on modern glaciers. One is that the mean altitude of the ice surface corresponds with the snow line; the other is that from one-half to one-third of the surface area of a glacier lies below snow line. These relations are reliable, however, only for small glaciers.

Calculations made on the basis of surface areas of the glaciers at the time of maximum advance during the Wisconsin stage gave approximate altitudes for snow line at that time ranging from 11,400 feet for the Platte and Middle Tarryall Glaciers to 11,800 feet for the Beaver Glacier, whereas calculations made on the basis of mean altitude of glaciers of the Wisconsin stage gave snow-line altitudes ranging from 11,450 feet for the Middle Tarryall Glacier to 12,000 feet for the Beaver Glacier. The maximum extent of the Beaver Glacier is less certain than of any other, and the Platte, Middle Tarryall, and Horseshoe Glacier are all too large for any great reliance to be placed on the calculations of their mean altitudes. In spite of all the uncertainties, however, it seems that the altitude of the snow line in Wisconsin time was somewhere between 11,500 and 11,800 feet. These figures are consistent with Capps' statement [8] that an altitude of 12,000 feet was sufficient at most places to start glaciation on the eastern slopes.

The Briscoe substage is represented by moraines only in the Platte, Sacramento, Horseshoe, and Middle Tarryall Valleys. Calculations of the snow line during this substage gave altitudes ranging from 11,900 to 12,100 feet by the surface-area method, and from 11,800 to 12,200 feet by the mean-altitude method; hence, the snow line at that time probably lay a little above 12,000 feet.

Uncertainties about the positions of ice borders for some distance above the moraines of the Alma substage make snow-line calculations for that substage impossible.

The only practical value of these calculations is the additional evidence they furnish for correlations between moraines of different valleys. This evidence, it should be stated, really is less positive than mere comparison of the figures might suggest.

[8] Capps, S. R., Jr., op. cit., p. 9.

PRE-WISCONSIN STAGES

The existence of a glacial stage in this region very much earlier than that represented by well-preserved Wisconsin moraines is proved by older moraines (see plate 10) found in both South Platte and Tarryall Valleys. Wide channels cut by streams in the older moraine contrast markedly with the narrow gorges in the younger; for example, near Fairplay the South Platte River had cut a channel 1,800 feet wide in the older moraine, but only 200 feet wide in the younger moraine. The entire cut in the older moraine was made during the time between maximum ice advance of the early glacial stage and maximum advance of the later stage, for the present stream is confined to a channel cut in outwash materials of the later stage. The interglacial stage was obviously several times as long, as the interval since the latest glaciation. For this reason it seems much more logical to assign the older moraines to a pre-Wisconsin glacial stage, as was done by Capps,[9] than to assign it to an early substage of the Wisconsin stage, as was done by Ray.[10]

The antiquity of the earlier glacial cycle is proved in other ways. In places its moraine has been eroded so much that only scattered boulders now remain; in fact, at several places all traces of moraine are gone, though swamps or hummocky bedrock testify to its former presence. Capps [11] mentions as additional evidence the weathered condition of the boulders. The writer confirms Capps' observation that, in general, boulders of the older moraine are more weathered than those of the younger moraine; nevertheless, fresh and weathered boulders may be found in either, and the differences are not sharp enough to serve as a reliable criterion to distinguish the moraines solely on this basis.

Ice retreat during the pre-Wisconsin glacial stage undoubtedly was interrupted by ice stands or readvances that deposited secondary moraines similar to those of the Wisconsin stage. An older moraine that lay within the area of the maximum advance of ice during the Wisconsin stage, however, has been swept away by the later ice.

As is shown in the detailed discussion of Tarryall Valley, there is some evidence suggesting a glacial stage in that valley still earlier than the two just mentioned.

Five complete stages, or cycles, of continental glaciation have been demonstrated for North America, and each stage must have had a counterpart in the mountains. It is not surprising, however, that the evidence of the earliest of these glacial advances is obscure. One can readily see that ice of a later stage sweeps away much of the earlier

---

[9] Capps, S. R., Jr., op. cit., p. 14.

[10] Ray, L. L., Glacial chronology of the southern Rocky Mountains: Geol. Soc. America Bull. 51, pp. 1901–05, 1940.

[11] Capps, S. R., Jr., op. cit., p. 14.

moraines in its path, and, consequently, that any glacial stage during which the ice did not advance farther than during all later stages is not likely to be represented by moraines. Furthermore, when the great difference in the amount of erosion to which the Wisconsin and pre-Wisconsin moraines have been subjected is considered it is quite conceivable that a correspondingly greater erosion of still older moraines may well have obliterated all traces of them, even though they were deposited outside the reach of later ice.

INTERGLACIAL STAGES

Little need be said here concerning the interglacial stages except to point out that the dominant processes in the valleys were rock weathering and stream erosion. Bedrock in the upper reaches was probably worn down slowly, whereas moraines farther downstream were easily dissected. Furthermore, the amount of material moved during interglacial intervals must have been much less than that moved by ice scour during the succeeding glacial stages. The results of interglacial erosion above the outermost of the Wisconsin moraines have no practical significance, and its results farther downstream are discussed in connection with the physiography of the South Park Province.

SOUTH PLATTE GLACIERS

WISCONSIN STAGE OF SOUTH PLATTE GLACIER

Maximum extent of ice.—The maximum advance of the South Platte Glacier during the latest or Wisconsin stage is clearly indicated by a terminal moraine—here called the Fairplay moraine—located half a mile upstream from Fairplay. As pointed out by Capps,[12] the ice during the Wisconsin stage extended from the Continental Divide at the head of Platte Gulch continuously down that stream for 15 miles and was joined by tributaries, the largest of which came from Buckskin, Mosquito, and Sacramento Gulches. The area of that glacier and its tributaries was 59 square miles.

Plate 9 shows the approximate area covered by ice at the time of maximum advance. Moraine clearly delineates the boundary throughout the lower part of the glacier system, except at a few places where ice lay against older moraine. Farther up the gulches, where erosion predominated over deposition, the position of the ice can at most places be determined by the presence of sheer cliffs that were developed along its sides. The outline indicated on plate 9 is essentially the same as that shown by Capps,[13] but two corrections to his mapping should be noted. One correction is that the South Platte Glacier overflowed into lower Beaver Creek, as is indicated by moraine that covers not only

[12] Capps, S. R., Jr., op. cit., pp. 80–85.
[13] Capps, S. R., Jr., op. cit., pl. 1.

the crest of the divide between the creeks but also its eastern slope down to Beaver Creek (see pl. 9). That the moraine came from South Platte Valley and not from the head of Beaver Creek is proved by its topographic continuity with moraine within the South Platte Valley, by its occurrence downstream below a long, unglaciated stretch of Beaver Creek, and by the abundance of pre-Cambrian boulders it contains. That it is of the Wisconsin rather than an earlier stage is less certain, yet its form and character, especially the abundance of boulders lying on the surface, contrast decidedly with the known "older" moraine, east of the same divide near Fairplay, which is represented by scattered boulders. The other correction to Capps' map is in the joining of the cliff glaciers on Mount Bross as tributaries to the South Platte. This correction was made because moraine was found on the east side of South Platte Valley near Hoosier Pass (see pl. 10) and on the west side along Windy Ridge considerably higher than the upper limit of ice shown by Capps. The requisite heightening of the ice boundary of the main glacier to approximately 12,000 feet (an altitude that Capps states is sufficient to start glaciation at most places along the eastern slopes of the ranges) makes it probable that the cliff glaciers joined the main ice body.

Stillstands and readvances.—The secondary terminal and lateral moraines and ice-border channels or both, that mark positions of the ice front during successive prolonged ice stands or readvances of the South Platte Glacier after it had begun its general retreat are shown on plate 9. From the standpoint of regional correlation as well as of economic interest the ice stands represented by the Briscoe and Alma moraines have the greatest importance, and, consequently, have been distinguished on plate 9. All the others, however, hold some economic interest, as will be discussed in the section on placer deposits.

The Briscoe moraine consists of well-defined terminal moraine that crosses the valley and connects on each side with lateral moraines bounded by conspicuous ice-border channels. The Alma moraine, by contrast, forms a low, smoothly rounded hill nearly a mile long and parallel with the valley; its east border is a wide but shallow depression that deepens southward and eventually curves around the south end as an ice-border channel; the west border is ill-defined. Two ice stands between the Fairplay and Briscoe moraines are represented by ice-border channels and associated lateral moraines, but not by a terminal arc; one ice stand between the Briscoe and Alma moraines is represented by a morainal ridge somewhat like that at Alma; four ice stands within 2 miles north of Alma and another 5 miles north of Alma are represented by typical terminal moraines; and a possible minor ice stand 4 miles north of Alma is represented by a poorly defined ice-border channel.

WISCONSIN STAGE OF BUCKSKIN GLACIER

The oldest terminal moraine in Buckskin Gulch is contemporaneous with the moraine at Alma, for during earlier substages the Buckskin Glacier coalesced with the Mosquito Glacier, which in turn joined the South Platte. Plate 9 shows the positions of the ice front during the Alma substage and during later stillstands.

WISCONSIN STAGE OF MOSQUITO GLACIER

Ice from Mosquito Gulch was tributary to the South Platte Glacier during the Fairplay and Briscoe substages. Terminal moraine representing the Alma substage forms a low but conspicuous belt of hummocks that crosses Mosquito Creek immediately above its mouth and joins lateral lobes on each side. Upstream from the ice front of the Alma substage there is evidence for at least four later stillstands, as shown by plate 9.

WISCONSIN STAGE OF SACRAMENTO GLACIER

Sacramento Glacier joined the South Platte Glacier at the time of maximum advance of ice during the Wisconsin stage but separated from it almost immediately after retreat began; consequently, a wide area in the lower part of Sacramento Gulch is now covered by thick moraine derived from both glaciers. This moraine is transected only by a narrow V-shaped gorge cut by Sacramento Creek, but it opens upstream, in the eastern part of sec. 36, T. 9 S., R. 78 W., into a large basin formerly occupied by a lake of impounded glacial water.

Position of the ice margin at the time of maximum advance during the Wisconsin stage is shown by plate 9; on both sides of the lower part of the valley, ice spilled over the bedrock divides. Plate 9 also shows the positions of the ice front during the Briscoe and Alma substages as well as during other determinable ice stands. Some are represented by secondary terminal moraines but the majority only by ice-border channels. No attempt was made to locate stillstands of post-Alma age in the Little Sacramento Valley.

MORAINES OF PRE-WISCONSIN AGE

Moraines deposited during an early glacial stage, probably the stage immediately prior to the Wisconsin, are found on both sides of the South Platte Valley, near Fairplay. The moraine on the west side forms a broad, low ridge between its own ice-border channel to the southwest and the ice-border channel of the younger moraine to the northeast. The ridge, which slopes gently toward the valley, apparently is everywhere covered with thick glacial debris, though the absence of natural or artificial cuts makes depth to bedrock unknown.

The glacial debris on the east side of the valley remains thick enough over most of the area to have prevented red coloration—a stain from the underlying rocks—from reaching the surface. However, in places the cover has been worn so thin that only scattered boulders rest on red soil, and elsewhere bedrock is exposed in cuts along the highway and the abandoned railroad line. Wherever the contact is exposed the rock surface is extremely irregular, with sharp ridges and depressions.

These two ridges are the lower parts of lateral moraines deposited by combined glaciers from the South Platte and Sacramento Valleys. They continue upstream to the south limit of younger moraine. Originally they must have continued downstream and joined to form a terminal moraine across the South Platte Valley nearly a mile below Fairplay, but during ice retreat and the subsequent interglacial stage, stream erosion removed the top of the terminal lobe. The base of the ancient terminal lobe and probably also some of the outwash apron now underlie outwash gravels of the latest stage.

The last vestige of an eroded moraine lying well above the edge of the Wisconsin moraines is represented by scattered boulders of pre-Cambrian rock that lie on top of and along the western and southern slopes of Bald Hill, south of Mosquito Creek. These boulders could have been deposited only by ice.

Two miles southeast of Bald Hill, where lateral moraines from the South Platte and Sacramento join, irregular hummocks in bedrock just outside the Wisconsin moraine suggest the possibility of a former moraine now eroded.

A fourth area of moraine that perhaps belongs to an early glacial stage exists in sec. 17, T. 9 S., R. 77 W. An east-to-west section across this moraine shows (1) an elongate area, about 1,000 feet wide, covered by thin glacial debris, or locally only by scattered boulders, (2) a prominent ice-border channel, and (3) thick glacial debris. These relations suggest that the channel represents the boundary of the maximum advance of the glacier during Wisconsin time.

FOURMILE GLACIER

WISCONSIN STAGE

At the time of maximum advance during the last, or Wisconsin, glacial stage, ice in Fourmile Gulch was 9 miles long and covered an area of 10½ square miles. Several of the cirques at the head of this gulch are particularly well developed. (See pl. 14,*A*). As shown on plate 9, the lower part of the glacier not only filled the valley but also spilled over the divides on each side and deposited moraine on their outer slopes. For a distance of half a mile along the northeast side, the overflow from Fourmile Gulch coalesced with the overflow from Sacramento Gulch.

The former position of the ice front is marked not by a terminal ridge but by moraine that merges topographically with the outwash apron to the southeast. Behind it, the entire valley for a distance of more than a mile is clogged with glacial debris, the result of the crowding and merging of lateral moraines rather than the deposition of true terminal moraine. Through this material the present stream has cut a winding, narrow gorge that has not yet reached the underlying bedrock. Sharp-crested ridges on each side of the valley undoubtedly represent preglacial bedrock divides whose crests and slopes have been covered by hummocky lateral moraine; each is bordered by a conspicuous ice-border channel near the terminus of the glacier. About a mile upstream from the terminus the heavy moraine forks within the valley and gives way gradually to a well-defined though narrow basin formerly occupied by a lake.

The Briscoe substage (see pl. 9) is represented by a terminal ridge nearly 50 feet high, which curves sharply on each side of the valley into a lateral lobe bounded by an ice-border channel. This moraine opens upstream into a well-defined basin. Moraine of the Alma substage, on the contrary, does not possess a terminal ridge and is separated only by ice-border channels from the next younger moraine. It is localized by a belt of relatively resistant rocks—Mississippian and Devonian dolomites—which lie at shallow depth beneath the surface. Five ice stands younger than the Alma substage are shown on plate 9. The three earliest are represented merely by ice-border channels within an area of moraine extending continuously upstream from the moraine of the Alma substage.

PRE-WISCONSIN STAGES

Although Fourmile Gulch undoubtedly underwent the same pre-Wisconsin glaciation as the South Platte, no trace of older moraine has been found there. A roughly circular basin located just southeast of the lowermost Wisconsin moraine may possibly be a former lake basin once bounded by moraine, but its identity is too doubtful to warrant any record on the map.

TARRYALL GLACIERS

WISCONSIN STAGE OF MIDDLE TARRYALL GLACIER

Maximum advance.—The glacial system of Middle Tarryall Gulch received its chief contribution of ice from the head of Montgomery Gulch, where the deep valley is bordered by precipitous cliffs, which are particularly high and rugged along the north face of Mount Silverheels. Ice not only filled the valley but also covered part of the

Beaver-Tarryall divide. (See pl. 9). At a point almost due north of Mount Silverheels the ice swept over the divide north of Montgomery Gulch and coalesced with ice from two relatively small cirques to the north. This broad sheet continued eastward until it encountered the peak located in sec. 10, T. 8 S., R. 77 W., which, as indicated on plate 9, formed a low island. Ice that flowed north of the peak perhaps was joined by tributary ice from the northernmost tributary of Middle Tarryall Creek, before rejoining the main glacier of Montgomery Gulch. Near this junction the united glacier spilled over the divide to the northeast and flowed down the outer slope as far as North Tarryall Creek. The entire sheet then continued southeastward to the present junction of North Tarryall and Middle Tarryall Creeks.

The position of maximum ice advance during the Wisconsin stage (pls. 9 and 10) is clearly shown by conspicuous moraine along the front and sides. Near the terminus, lateral moraines form clear-cut ridges. These lateral ridges curve into terminal debris, which rises rather abruptly some 50 feet above outwash materials in the stream.

Less than half a mile upstream from the former ice terminus, the frontal moraine opens into a well-developed interior basin, which extends to the moraine of the Briscoe substage. Within this interval are suggestions of two ice-border channels indicating ice stands similar to those of the South Platte Valley. The interior bench of the Fairplay substage is here well developed only along the lateral moraine of the northeast side.

Stillstands and readvances.—Termination of the Middle Tarryall Glacier during the Briscoe substage, shown on plate 9, is best seen north of the creek where obscure terminal moraine within the valley grades into a sharp-crested lateral ridge bounded by a wide and deep ice-border channel. Another ice-border channel located a quarter of a mile farther upstream is doubtless to be correlated with the inner arc of the Briscoe moraine in the South Platte Valley.

During the Alma substage the two tributary glaciers of Middle Tarryall Gulch did not join. The former ice front in Montgomery Gulch (pl. 9) is revealed by abundant moraine, which merges upstream with slightly younger moraine. Evidence for at least four ice stands later than the Alma substage may be seen in Montgomery Gulch. The ice front in Deadwood Gulch during the Alma substage is somewhat obscure but has been drawn on plate 9 where the valley profile changes from U-shaped to V-shaped.

WISCONSIN STAGE OF NORTH TARRYALL GLACIER

Existence of a small glacier in the upper part of North Tarryall Gulch is demonstrated by an area of moraine (pl. 10) that terminates three-quarters of a mile above Selkirk Creek. This termination must represent the farthest advance during the Wisconsin stage, for no additional moraine, except that deposited by the Middle Tarryall Glacier, is found downstream.

Ice scour above this moraine has at most places been so slight that the outline of the former glacier cannot be sketched with any assurance. Absence of clear-cut cirques suggests that the collecting ground along the Continental Divide and the west face of Boreas Mountain was but thinly covered by snow and ice. If a glacier existed in North Tarryall Valley during the Briscoe and Alma substages, it was too small and weak to form definite moraines; indeed, the altitude of most of the headward basin of this valley was probably too low to have initiated glaciation during those substages.

PRE-WISCONSIN STAGES

A moraine lying outside the limits of the Wisconsin ice sheet, and therefore older, is found along the south bank of Tarryall Creek in secs. 19 and 20, T. 8 S., R. 76 W. It is thickest and best exposed close to the stream in section 20, where it closely resembles the younger moraine except in containing many boulders of Dakota sandstone. Both uphill and westward from its thickest point, this older moraine thins until near its margins it is represented only by scattered boulders. Immediately above the highest vestiges of moraine in sec. 19 the ground is swampy.

Additional evidence of an earlier glaciation is furnished by a small area of hummocky ground near the corner between secs. 2, 3, 10, and 11,T. 8 S., R. 77 W. There the low hummocks are composed entirely of bedrock, but their configuration strongly suggests that moraine, now eroded, once overlay them.

More obscure but similar evidence of a lateral moraine is found along the east slope of the ridge trending southeastward across sec. 29, T. 8 S., R. 76 W. In the northwest part of sec. 20, T. 8 S., R. 76 W., two rounded knolls of porphyry bedrock that project above the outwash apron may also be vestiges of a former terminal moraine.

The evidence in the first two of these areas suggests that during this early glacial stage, immediately preceding the Wisconsin, ice from Selkirk Gulch and North Tarryall Creek coalesced with ice from Middle Tarryall Gulch to extend down the valley some 2 miles beyond the Wisconsin ice front. If the obscure topographic features of the last two areas named really represent an ice advance, this advance may have taken place during an even earlier stage.

880373—50——3

MINOR GLACIERS

BEAVER GLACIER

Glaciation in the upper part of Beaver Creek is proved by the broad cirque at the head of the valley, the hanging tributary valley on the southeast slope of Mount Silverheels, and the moraine (see pl. 10) in the vicinity of that tributary. Even where best developed the moraine forms only a thin cover, and at places it is represented only by scattered boulders. Moreover, the main valley above the moraine is essentially V-shaped and only slightly wider than if cut by stream action alone. These features could be interpreted as indicating either (1) that the moraine marks the maximum advance of ice during a stage preceding the Wisconsin and that the original effects of glaciation have since been modified by a comparatively long period of stream erosion, or (2) that the Wisconsin ice sheet was so small and weak that it produced little modification of the valley profile and deposited only scant moraine. The latter hypothesis is here favored, and the moraine has been mapped in plate 10 as of Wisconsin age.

TROUT CREEK GLACIER

Hummocky moraine in upper Trout Creek west of Little Baldy Mountain marks the maximum advance, during the Wisconsin stage, of ice that accumulated in small cirques on the southeast side of Mount Silverheels. The outline of this moraine (see pl. 10) was sketched from Palmer Peak.

Scattered porphyry boulders and the U-form of a saddle between Trout Creek and the stream that heads in sec. 2, T. 9 S., R. 77 W., suggest that ice during an earlier stage reached or spilled over this saddle. The swampy terrain in the northern part of sec. 1 of the same township also suggests the former presence of ice.

CROOKED CREEK GLACIER

A small ice sheet accumulated in the upper part of a narrow valley bordering the west face of Palmer Peak and deposited a very small moraine (pl. 10) in the saddle a mile south of the peak. Ice scour by this glacier was too feeble to leave evidences by which an accurate delineation of the area covered could be made; so the ice boundary shown on plate 9 must be regarded as conjectural.

MATERIALS IN THE MORAINES

Moraines deposited by the South Platte Glacier and its tributaries and by the Fourmile Glacier comprise a heterogeneous assemblage of pebbles and boulders derived from all the types of rocks cropping out upstream from them: Pre-Cambrian granites, gneisses, and pegmatites; Paleozoic sedimentary rocks ranging from the Sawatch quartzite to the lower division of the Pennsylvanian and Permian sequence; and Tertiary igneous rocks of several types. Most abundant are pre-Cambrian granite and strongly granitized gneiss and Tertiary porphyries. Quartzites, less granitized facies of gneiss, and pegmatite are fairly common, however, and even dolomite and less resistant types may be found. The pre-Cambrian boulders are particiularly important, as they serve to distinguish materials derived from west of the Platte-Beaver Divide from those derived east of it.

The most resistant rocks above the belt of Dakota sandstone in Tarryall Valley are the widespread Tertiary igneous rocks, which therefore constitute the most abundant boulders and pebbles of moraines in that drainage area. They are associated, however, with red sedimentary rocks, especially sandstones and conglomerates. Green contact-metamorphosed rocks, which are more durable than the unmetamorphosed red beds, are abundant in all moraines deposited by ice from Middle Tarreyall Gulch but are nearly absent in moraines of the North Tarryall glacier. Moraines deposited downstream from the outcrop of the Morrison and Dakota formations of course contain an abundance of sandstone from these formations.

Traces of gold may be found in all moraines deposited by ice that crossed one or more of the lode-gold areas indicated on plate 8. Commercial placers, on the other hand, are restricted to certain favorable places, as will be discussed in the section on placers.

PHYSIOGRAPHY OF THE MOUNTAIN-PARK PROVINCE

OUTLINE OF PROCESSES AND EVENTS

The topographic form of the deposits and the distribution of placer gold today in the northern part of South Park result from physiographic evolution that began soon after the Laramide revolution. The dominant processes have been stream erosion and deposition, but the character of the bedrock has greatly influenced the results. A particularly noteworthy feature is the prevalence of lateral erosion in certain areas where relatively soft rocks—Upper Cretaceous and Tertiary in the eastern part of the area, and Pennsylvanian and Permian in the western part—crop out upstream from relatively resistant granite. Under these conditions, the rate at which streams can deepen their channels in granite determines gradients upstream, where the excess erosive power in the softer rocks can be expended only in lateral cutting while the channels in granite are being slowly deepened. The two main granite barriers that have affected physiography in the northern part of South Park constitute the range of hills along the east margin of the mapped area (pl. 8), and the ridge that is transected by the South Platte River near Hartsel, 17 miles south of Fairplay and 13 miles south of the area represented on plate 10.

The first event directly related to the concentration of placer gold occurred long before the glacial period, when ancestral streams issuing from the mountains attained a temporary base level. At that time the area included two vast river plains, one across the present drainage of the South Platte River and Fourmile Creek, the other across most of the present drainage of Taryall, Michigan, and Jefferson Creeks. These plains may be reconstructed from a few existing remnants, mapped as terrace No. 1 on plate 10. Each plain in detail consisted of several smooth-surfaced, nearly flat, coalescing fans which inclined gently outward from points where streams emerged from the mountains. A veneer of gravel ranging in thickness from a few feet to several tens of feet overlay the bedrock.

Owing either to more active erosion downstream or to uplift in the Park Ranges, these plains became dissected. Eventually, probably not much before the glacial period, another temporary base level was attained and another set of river plains was formed similar to the first set and nearly as extensive. A few remnants of the earlier plains now stood as terraces above the valley floors. The younger plains in turn were extensively dissected when the streams were rejuvenated a second time. Remnants of plains from the second base level are indicated on plate 10 as terrace No. 2.

During the glacial period, stream processes throughout most of northern South Park were controlled by ice advances and retreats within the mountains. Doubtless each ice advance was accompanied by deposition of outwash aprons and each retreat by stream erosion, but the results attained during earlier glacial stages have been obscured by those of the last two. Immediately prior to the last pre-Wisconsin glacial stage, streams in South Park were carving broad, shallow channels in places where soft bedrock or gravel permitted considerable lateral erosion. During the glacial stage that followed, these wide stream channels became areas of deposition of outwash gravels which grade downstream into more typical stream gravels. Renewed erosion during the last interglacial stage carved out new valley floors upon which outwash from the glaciers of the Wisconsin stage was in turn deposited.

Although the valley floors that existed during pre-Wisconsin glacial time have been extensively eroded, remnants preserved in places as low benches have been mapped on plate 10 either as pre-Wisconsin outwash aprons or as terrace No. 3, according to their distance from the ice front and the resultant character of the gravel. Valley floors of Wisconsin age, on the contrary, have been only slightly incised by the present streams, and outwash aprons of Wisconsin age are therefore almost perfectly preserved.

During the relatively short postglacial period streams have cut shallow channels, but other topographic changes have been insignificant. The amount of vertical deepening along each stream decreases gradually downstream from a maximum near the mountains, where it ranges in different valleys from a few feet to about 50 feet. As a consequence, the former valley floors of Wisconsin age remain as flood plains of present streams over much of the area, and Wisconsin outwash aprons merge downstream into alluvium. At a few places, one or two miniature terraces, or benches, separated vertically by only a few feet, can be distinguished and mapped between the stream bed and the Wisconsin surface. These benches can doubtless be correlated in age with the Briscoe and Alma substages.

SOUTH PLATTE RIVER DRAINAGE AREA

TOPOGRAPHIC FEATURES

The South Platte River, as shown by plate 10, is joined by Beaver and Crooked Creeks within the area here considered, and Crooked Creek itself is joined by several small tributaries. The eastern divide for this drainage is a bedrock ridge, rising some 500 feet above the streams, capped by resistant sandstones of the Morrison and Dakota formations. It stood above the valley floor throughout the preglacial and glacial periods.

In sec. 22, T. 10 S., R. 77 W., the western divide of the South Platte drainage system is a remnant of the earliest valley floor, mapped as terrace No. 1. This terrace (see pl. 14, B) extends southward about 10 miles beyond the area represented on plate 10. At its north edge it has been dissected, and so terminates as several rounded knolls beyond which is bedrock that formerly underlay the terrace.

Two areas in the southern part of T. 9 S., R. 77 W., having smooth, nearly plane surfaces have been mapped as terrace No. 2. This terrace is bordered in some places by a low escarpment leading down to the present streams and in other places by bedrock hills that rise above it. As the terrace stands at altitudes of several hundred feet below the northward projection of terrace No. 1, it clearly is younger than the oldest terrace of the region; however, its pre-glacial age is attested by the absence of pre-Cambrian boulders anywhere except at a dumping ground for the town of Fairplay. The implication is that the terrace already was dissected so as to stand above th valley floor when the first ice or glacial water loaded with pre-Cambrian boulders flowed down Beaver Creek.

Outwash aprons of pre-Wisconsin age are not exposed in the South Platte Valley but may lie buried beneath younger gravels downstream from the front of the pre-Wisconsin ice sheet, which was about a mile southeast of Fairplay.

A major part of the present valley floor is a remarkably smooth plain, slightly arched or fan-shaped in cross section, composed of outwash materials from the Wisconsin ice sheet. At Fairplay, this outwash apron forks, the main branch extending northward to the Fairplay moraine and the smaller branch to the former ice front in Beaver Creek. The two branches are separated by older moraine, as shown on plate 10. The valley plain extends about 10 miles to the south, nearly to Hartsel, but the surface gravels grade into typical stream gravels.

The South Platte River has cut the Wisconsin outwash plain to a depth of about 50 feet at Fairplay, but the depth decreases downstream and is less than 5 feet at the south border of the area shown on plate 10. Two temporary base levels, each of short duration, are represented by low benches above the river bed but below the Wisconsin outwash surface. Each bench is bordered by an escarpment that diminishes in height and eventually loses its identity southeastward.

Though devoid of economic interest, the complex distribution of bedrock, alluvium, and terrace in Crooked Creek Valley deserves brief explanation. The two areas of terrace No. 2 originally joined and extended northward to the mountains, but stream erosion during the glacial and postglacial periods destroyed most of the terrace and carved out a shallow basin in the underlying bedrock. As Crooked Creek could not cut deeper than a profile determined by the depth of the South Platte River, its excess power of erosion resulted in lateral cutting where rocks were soft, particularly in the basin north of the Fairplay-Denver highway. The floor of the basin later was covered with alluvium, above which low mounds and ridges composed of the more sandy strata project.

UNCONSOLIDATED DEPOSITS

Terrace No. 1 gravels.—Gravels covering terrace No. 1 as well as the outlying rounded peak to the north are considerably more quartzose than any other deposits in the area. Boulders up to 1 foot in diameter are scattered among smaller pebbles and grains, all of which are considerably rounded. White quartzite predominates in the gravels but is accompanied by quartzose facies of the nonred Pennsylvania beds, minor amounts of jasperoid and Tertiary porphyry, and a few boulders of pre-Cambrian rocks.

Terrace No. 2 gravels.—The boulders and pebbles in gravel overlying terrace No. 2, east of Beaver Creek, are predominantly of Tertiary porphyry but include considerable quantities of the more resistant facies from the red-bed series, and subordinate amounts of green contact-metamorphosed rocks. The largest boulders are more than a foot in diameter. Those of porphyry tend to be more rounded than those of sedimentary rock. The absence of pre-Cambrian rocks indicates that the material was derived entirely from the area east of the South Platte-Beaver Divide.

Outwash gravels.—Outwash gravels of Beaver and South Platte Valleys form a heterogeneous assemblage in which pre-Cambrian rocks and Tertiary porphyries greatly predominate. Quartzites and sandstones are conspicuous, and all the other sedimentary rocks may be found. All the boulders and pebbles are somewhat rounded. The largest are about 2 feet in diameter, but the size diminishes progressively downstream from the terminal moraine, so that at the south end of the area the largest boulders are less than a foot in diameter. Where exposed in placer cuts the material shows little evidence of sorting or stratification. It differs notably from moraine only in the absence of any huge boulders and in the tendency for flattish stones to be shingled.

Deposits of the two terrace levels east of the South Platte River near the Purcell ranch are identical in composition with the outwash gravels on the west side of that stream.

Other deposits.—The stream alluvium ranges in coarseness from fine-grained silt at some localities to gravel at others.

Talus on the bedrock slopes is of no economic interest in this report.

FOURMILE CREEK DRAINAGE AREA

TOPOGRAPHIC FEATURES

Fourmile Creek drains all the area east of the Mosquito Range that lies south of Sacramento Creek and west of the South Platte drainage. Near the south end of the outermost Wisconsin moraine the main creek turns abruptly eastward, cuts across the north tip of the adjoining outwash apron, passes through a narrow bedrock gorge, emerges northeastward into an open basin, bends and flows southeastward for about 2 miles, and finally turns southward. Thirteen miles southeast of the area represented on plate 10, Fourmile Creek joins the South Platte River.

South of Fairplay the area north of Fourmile Creek comprises a considerably dissected terrace on the west and a gently sloping hillside on the east. These two major features converge along a line trending northward from the point where the Fairplay-Buena Vista highway crosses Fourmile Creek. The western terrace lies inside an angle formed by the divide south of Sacramento Creek and northeast of upper Fourmile Creek. (See pl. 10.) Originally it was a smooth, gravel-covered plain that sloped very gently southeastward, was continuous with the terrace south of Fourmile Creek, and presumably extended for several miles farther south. Correlation of this terrace with others cannot be made with certainty. It is much lower in altitude and therefore younger than terrace No. 1, and it stands above and is therefore older than the Wisconsin outwash apron. Thus, it has at least approximately the same age as terrace No. 2 of South Platte Valley. As neither valley contains more than one terrace of this approximate age, this terrace is classed as terrace No. 2 on plate 10.

The gently sloping hillside on the east side of Fourmile Valley is contemporaneous in age with terrace No. 2 to the west. Its original gravel cover remains almost undisturbed because there have been no streams to dissect it. Slight indentations near the Hall ranch are due entirely to local runoff water.

Most of the terrain for several miles south of the easterly course of Fourmile Creek is covered with outwash from the last Fourmile Glacier. This outwash apron forms a broad low fan, whose axis slopes southeastward. To the east, the gentle slope of the outwash plain merges without topographic demarcation into the flood plain of lower Fourmile Creek. To the west it passes into the alluvium of High Creek. To the north, the outwash apron is separated from terrace No. 2 by an escarpment that is 20 feet high at the west end but only a foot high at the east end.

UNCONSOLIDATED DEPOSITS

Gravels on west slope of Fourmile-South Platte divide.—Gravels of the gently sloping hillside on the east side of Fourmile Valley contain rounded boulders up to a maximum of 1 foot in diameter, though very few are more than half that size. Most of the boulders and pebbles represent sandy facies of the nonred Pennsylvanian beds, but many are of porphyry, and a few are of red beds or pre-Cambrian rocks. Toward the south, as the terrace No. 1 is approached, appreciable amounts of white quartzite and jasperoid appear.

The scarce though widespread pre-Cambrian boulders originally came from the present South Platte drainage area. Their present location suggests that the South Platte River for a time flowed through a low saddle now traversed by the Fairplay-Buena Vista highway, though the boulders conceivably could be derived from terrace No. 1, which formerly covered the area.

Gravels of terrace No. 2.—The gravel covering terrace No. 2 contains moderately rounded boulders having a maximum diameter of 1½ feet. Boulders and pebbles of quartzite, conglomerate, arkose, and sandy shale from nonred Pennsylvania strata are abundant; those of porphyry are scarce, of red beds fewer, and of pre-Cambrian rocks still fewer in number. Locally, jasperoid and coarse-grained, angular quartz pebbles are intermixed.

Stratified deposits beneath gravels of terrac No. 2.—There are poorly cemented, well-stratified deposits underlying the surface gravel in cuts

A GLACIAL CIRQUES AT HEAD OF FOURMILE GULCH, LOOKING WEST.

B TERRACE NO. 1, EAST OF FOURMILE CREEK, LOOKING SOUTHEAST.

Shallow stream channel of Fourmile Creek in foreground is cut in glacial outwash apron. Ridge along skyline is Terrace No. 1 extending southward from margin of area shown on map (plate 10).

A TERRACE NO. 2 GRAVEL OF TROUT CREEK RESTING ON UNEVEN SURFACE OF BENTON SHALE.

The gravel is about 10 feet thick. Road cut along Fairplay-Denver highway, looking south.

B CLINE BENCH PLACER CUT.

Note layers of coarse gravel separated by thin layers of fine gravel which lens out to the right. Looking *southeast.*

of the Hall placer and also in a cut formed by leakage of an irrigation ditch southeast of the Hall ranch. Because they are exposed only in artificial cuts, they have not been indicated on plate 10. Layers less than a foot thick are composed of sediments which range in texture from silt to a gravel containing well-rounded pebbles up to several inches in diameter. Color ranges from white to yellow-brown. If these deposits are merely local, they can be interpreted as fillings of depressions locally formed below the general level of the pediment on one side of the valley and the sloping hillside on the other. If, however, they prove to be of wide extent, they must be regarded as indicating a cycle of erosion and deposition intervening between the cycles of the No. 1 and No. 2 terraces.

Gravels of outwash apron south of Fourmile Creek.—Gravels of the outwash apron south of Fourmile Creek contain fairly well-rounded boulders and pebbles. They are mostly less than 6 inches but some exposed along the Fairplay-Buena Vista highway are 1 foot in diameter. Closer to the source, their maximum diameter is 2 feet. The materials consist of a heterogeneous mixture of early Paleozoic dolomite and quartzite; Pennsylvanian nonred conglomerate, quartzite, arkose, and sandy shale; Tertiary porphyry; pre-Cambrian rocks; a little jasperoid; and, rarely, red beds. Pre-Cambrian boulders are considerably more abundant than in any of the earlier deposits, but they were derived from upper Fourmile Gulch.

Alluvium.—Fine-grained mud and silt that cover depressions at various localities and also flood-plain deposits along the streams have been mapped as alluvium.

VALLEY EAST OF THE SOUTH PLATTE

A curious valley, the upper part of which is much too deep to have been carved out by its present intermittent stream, occupies a belt of soft Benton shale east of the escarpment formed by rocks of the Morrison and Dakota formation; this valley represents the original course of Trout Creek. During an early glacial stage Trout Creek became diverted into the Park Gulch drainage basin; later, probably during the interglacial stage just preceding the Wisconsin glaciation, headward erosion of its former, deeper valley recaptured Trout Creek but at a different place from its former point of diversion.

At most places the bottom of the valley is bordered by fairly steep walls, not more than 25 feet high, which grade upward into less steep slopes. Several remnants of a low bench or terrace and several alluvium-covered depressions in the bedrock along this valley have been indicated on plate 10.

DRAINAGE AREA OF TARRYALL, MICHIGAN, AND JEFFERSON CREEKS

TOPOGRAPHIC FEATURES

General relations and boundaries.—The physiographic history of Tarryall, Michigan, and Jefferson Creeks is so closely interwoven that their entire drainage area can most conveniently be considered as a unit. These three creeks, as well as all their larger tributaries except Park Gulch, originate in the mountains that carry the Continental Divide from the Front Range across to the Mosquito Range, and their upper courses have therefore been glaciated. The results of glaciation in Tarryall Creek have already been discussed; a description of the glaciation in Michigan and Jefferson Creeks is not included in this study because it has no bearing on the distribution of placer gold.

Park Gulch, a tributary of Tarryall Creek, consists of three main branches which head either on the southeast slope of Little Baldy Mountain or in the lowland to the east, north and south of Como. Trout Creek above its point of recent capture by an arm of the next valley to the west is included because it formerly joined the South Branch of Park Gulch. The southwest boundary of the area considered in this section of the report, therefore, is the Trout Creek watershed as it was during the pre-Wisconsin glacial stage, and not the present watershed.

Mapping was extended northeastward as far as Jefferson Creek, or a little beyond, well outside the limit of terrain having any interest to prospectors.

Terrace No. 1.—Seven remnants of the earliest temporary base level in this drainage area have been mapped as terrace No. 1. All are within a mile of Tarryall Creek (see plate 10). The largest extends for more than 2 miles southeastward from sec. 25, T. 8 S., R. 76 W.

Terrace No. 2.—The second temporary base level of erosion, mapped as terrace No. 2 (see pl. 10), is extensively represented. The westernmost exposure is west of Trout Creek, where a small remnant slopes southeastward from a bedrock knoll. If its surface slope were projected eastward across Trout Creek it would join another remnant crossed by the Fairplay-Denver highway. This terrace level can readily be projected across the mile gap to a much larger remnant between middle Park Gulch and its south branch, and this remnant in turn is topographically continuous with the large terrace extending from a point a mile and a half north of Como to the granite hills along the east border of the area.

A former extension of the terrace north of Tarryall Creek is indicated by several remnants along the divide between Packer Gulch and Tarryall Creek. All but one of them are shown on plate 10 without definite boundary because they are mere vestiges on a much dissected ridge. A little more erosion would have destroyed even these vestiges, whereas a little less would have left them united as a continuous surface.

Several remnants of a bench along the north face of the granite ridge between Packer Gulch and Michigan Creek lie at the projected level of terrace No. 2. They indicate that the Packer-Michigan Divide, unlike the Packer-Tarryall Divide, stood above the terrace level, undoubtedly because of the resistant character of the bedrock. The benches however, indicate that the terrace did extend across the present Michigan Creek to the broad granite ridge north of the Fairplay-Denver highway.

Where terrace No. 2 has been dissected, abundant outcrops of bedrock, or fragments of bedrock around gopher holes, show that the terrace is a bedrock pediment covered with gravel. The bedrock surface is irregular, as shown by pl. 15, *A*. In places the gravel cover is less than a foot thick, though elsewhere it is more than 10 feet thick.

The age of terrace No. 2 can best be determined from its relations to the moraines of the Tarryall glaciers. It quite clearly antedates the Wisconsin glaciation, for Wisconsin outwash materials cover the floors of valleys cut into the terrace. Lack of topographic continuity of the terrace northwest of Como indicates that it was somewhat eroded at the time of deposition of the pre-Wisconsin glacial outwash. Thus, it may reasonably be correlated with terrace No. 2 mapped in the South Platte and Fourmile Valleys.

Terrace No. 3.—The present stream pattern was developed after terrace No. 2 began to be dissected. All the steps of development, however, cannot be followed in detail. Evidence furnished by wind gaps at the head of Packer Gulch, in the divide between Packer Gulch and Tarryall Creek, in the northern part of sec. 26, T. 8 S., R. 76 W., and between Tarryall and Michigan Creeks within the eastern range of granite hills testify to a great many minor changes in drainage, as do water gaps at the point where Tarryall Creek flows through the relatively narrow gorge cut in rocks of the Laramie (?) formation in sec. 26, T. 8 S., R. 76 W., and where Michigan Creek flows eastward from Michigan Station. The general condition doubtless was one of streams emerging from the mountains into broad valleys and forming braided channels, with the main flowage now in one, now in another.

A temporary erosion level of relatively short duration, younger than Terrace No. 2, has been preserved by benches that locally cover sufficient area to be called terraces. Where well developed they have been mapped as terrace No. 3. (See pl. 10.) They were never as extensive or as uniformly flat, however, as the earlier terraces.

Remnants of a low terrace are seen along both banks of Jefferson Creek, southeastward from the town of Jefferson, and along the Jefferson-Michigan Divide in T. 7 S., R. 76 W. If these remnants are the same age, as seems highly probable, terrace No. 3 formerly covered the present Jefferson Creek Valley continuously from the eastern hills to the foot of the mountains, where it also extended across to and down Michigan Creek.

Along Tarryall Creek benches of the low terrace are most extensively preserved in the vicinity of Milligan Lakes, but a small remnant a mile to the southwest indicates that the terrace probably once covered the entire basin from Milligan Lakes to the northern part of sec. 8, T. 9 S., R. 75 W. It covered a much smaller area, however, than the low terrace of Jefferson Creek. Immediately preceding maximum development of this terrace, Tarryall Creek probably flowed through the wind gap east of Milligan Lakes.

The considerable difference in altitude between terrace No. 3 of Jefferson Creek and terrace No. 3 of Tarryall Creek can be interpreted as meaning either (1) that the two terraces are contemporaneous in age and that Jefferson Creek had cut to a lower level at the time, or (2) that the Tarryall terrace is the older. The former interpretation is here preferred because Jefferson Creek now has a lower elevation than Tarryall Creek where they enter the eastern granite hills, and because the terraces have approximately the same height above the present valley bottoms.

Along the South Branch of Park Gulch, terrace No. 3, as indicated on plate 10, is well-developed west of the gap through the north-south ridge composed of rocks of the Laramie (?) formation. Here it extends northward far enough to establish continuity with a pre-Wisconsin outwash apron that forms the broad plain between the Fairplay-Denver highway and the southeast base of Little Baldy Mountain. Benches also occur along the lower courses of all branches of Park Gulch but are too imperfectly preserved to warrant representation on the map.

Outwash aprons of pre-Wisconsin age.—Three remnants of an outwash apron derived from a pre-Wisconsin glacier of Tarryall Creek have been mapped in T. 8 S., R. 76 W. They presumably are contemporaneous with terrace No. 3 farther downstream. One of the outwash remnants, locally called the Cline Bench, has been the scene of commercial placer operations. (See pl. 10.) Bedrock underlying the bench is covered by gravel which in places has a thickness of as much as 36 feet. The age of the Cline Bench is proved by its relative altitude—below terrace No. 2 and above the outwash apron of Wisconsin age.

At the head of the South Branch of Park Gulch is a broad plain whose surface materials and whose topographic relations to the postulated ice front indicate an outwash apron from a glacier of pre-Wisconsin age in Trout Creek Basin. The outwash apron is bordered on the west by a bench here regarded as of Wisconsin age, which in turn is cut by the present stream gorge of Trout Creek.

Outwash aprons of Wisconsin stage.—An outwash apron of Wisconsin age covers Jefferson Valley for more than a mile below its emergence from the mountains except for a narrow stream channel. As the stream channel becomes more shallow to the south the apron merges without topographic break into flood-plain materials.

In Michigan Valley, glacial outwash below the lowermost Wisconsin moraine is largely covered with recent debris from Antelope Gulch and several other small creeks farther south.

The outwash apron within the narrow channel of Tarryall Creek near Peabody's Switch has been largely eroded. Where it emerges from the mountains, however, it forms a low fan-shaped plain that forks against the Cline Bench outlier in sec. 21, T. 8 S., R. 76 W. The northern fork extends as far as Michigan Creek; the southern fork extends eastward, one part swinging around the Cline Bench terrace to join the northern fork, and the other part merging with the flood-plain materials of present Tarryall Creek. Plate 10 shows a stream-cut bench in the outwash apron north of Tarryall Creek. This bench is scarcely perceptible, for it stands only 2 or 3 feet below the original level of the outwash apron.

Depth to bedrock over part of the Tarryall outwash apron is shown by plate 13, where the gravel cover is much thinner than in the South Platte Valley.

Miscellaneous features.—Areas mapped as terrace-border slopes at most places are relatively steep escarpments, but they have a gentle inclination in the lower reaches of Park Gulch. The many bedrock exposures on them, and the still more numerous places where bedrock has been brought up from deep gopher holes indicate that bedrock everywhere underlies the terraces at no great depth.

Irregularly dissected areas other than terrace-border slopes are mapped as bedrock hills. Four of them—the eastern granite hills, the northward-trending ridge composed of beds of the Laramie (?) formation and associated igneous rocks, the area of volcanic rocks west of Park Gulch, and a part of the Packer-Michigan Divide—stand above terrace No. 1 and so must have been persistent watersheds throughout the evolution of South Park. The large area between Michigan and Jefferson Creeks north of the Fairplay-Denver highway is too remote from any remnants of terrace No. 1 or terrace No. 2 to permit determination with certainty whether it now stands above or below the projected terrace levels; however it was probably once covered by these terraces.

South of Park Gulch in T. 9 S., R. 76 W., is an area mapped as one of the miscellaneous debris-covered hillsides. It is part of a valley that heads some 2 miles south of the area mapped. The present irregularity of the floor of the valley is due to dissection of a preexisting pediment of terrace No. 3 age but is too steep to be mapped as a part of that terrace.

More than a dozen roughly circular depressions, ranging from 100 to several hundred feet in diameter and no more than 30 feet deep—and thus too small to be mapped—are scattered irregularly over the terrain south of latitude 39°20′. Most, if not all, have been excavated through the gravel covers of the terraces and into the bedrock. They are topographic curiosities of unknown origin, perhaps excavated by wind.

UNCONSOLIDATED DEPOSITS

Gravel capping terrace No. 1.—Gravel capping the six remnants of terrace No. 1 closest to Tarryall Creek is derived from the upper reaches of that creek. Some of the pebbles, cobbles, and boulders are well rounded and others only slightly so. Few cobbles exceed 6 inches in diameter, but there are a few as much as a foot in diameter. Pebbles of Dakota sandstone predominate, those of porphyry are abundant, and those of red beds, green contact-metamorphosed rock, and Benton shale are present in minor amounts.

A peak in sec. 19, T. 8 S., R. 75 W., mapped as terrace No. 1, is now only a bedrock vestige of a former terrace from which all gravel has been eroded.

Gravel capping terrace No. 2.—Trout Creek furnished the gravel capping two remnants of terrace No. 2 in sec. 6, 7, and 8, T. 9 S., R. 76 W. The materials are poorly rounded and range up to more than a foot in diameter. Porphyry pebbles predominate, those of Dakota sandstone and red shale are subordinate in amount, and those of Benton shale and green contact-metamorphosed rock are very minor constituents. The gravel probably does not exceed 10 feet in thickness at most places. The underlying bedrock surface, as seen in a highway cut, is in places smooth, in others very ragged. (See pl. 15, *A*.)

All the remainder of terrace No. 2 is capped by gold-bearing gravel from Tarryall Creek, the boulders, cobbles, and pebbles being all more or less rounded. Their size ranges from a maximum of nearly 2 feet in longest dimension, north of Como, to 6 inches near the southeast corner of the area. Boulders exceeding 1 foot in diameter are found as much as 2 miles southeast of Como, yet nowhere are they abundant and all of them are of Dakota sandstone. The other pebbles consist chiefly of Dakota sandstone and of porphyry, with subordinate amounts of red beds and green contact-metamorphosed rock and a very few of Benton shale. Much of the porphyry is moderately weathered, and some is rather rotten. The most striking difference between the gravel from Tarryall and that from Trout Creek is the greater quantity of green contact-metamorphosed rock in the former.

No gravel remains on benches representing terrace No. 2 level between Packer and Michigan Creeks.

Gravel capping terrace No. 3.—Gravel covering the low terrace along the left bank of the South Branch of Park Gulch is probably a mixture of early outwash materials from Trout Creek and reworked gravel from terrace No. 2, which formerly extended across the gulch. Boulders are moderately rounded, do not exceed a foot in maximum diameter, and diminish gradually in size southeastward. The coarser constituents are predominantly Dakota sandstone and porphyry but include minor quantities of red beds, green contact-metamorphosed rock, and Benton shale.

Terrace No. 3 of lower Tarryall Creek and also unmapped benches in lower Park Gulch are capped with gravel composed of granite and pegmatite similar to the bedrock to the north and east. The cobbles and pebbles, all less than 6 inches in diameter, are rather well-rounded. Locally, the pre-Cambrian materials are mixed with materials derived from terrace No. 2.

Terrace No. 3 along Jefferson Creek is capped with gravel that includes pre-Cambian granite, granitized schist, and pegmatite, with subordinate amounts of porphyry, Benton shale, and Dakota sandstone, all of which were derived from the mountains. The maximum size of boulders diminishes southeastward and few found southeast of the town of Jefferson exceed 6 inches in diameter.

Cline Bench gravel.—The gravels of the Cline Bench contain mainly porphyry with slightly less abundant Dakota sandstone and with considerable though subordinate amounts of red beds and green contact-metamorphosed rock. Much of the porphyry is considerably weathered. Three layers can easily be distinguished. (See pl. 15, *B.*)

Tarryall outwash apron from Wisconsin stage of glaciation.—The largest boulders of the Tarryall outwash apron are mostly of porphyry with a subordinate number of Dakota sandstone. From a maximum diameter of more than 2 feet where the stream emerges from the mountains, they become progressively smaller downstream. Among the smaller cobbles and pebbles of porphyry and sandstone is a considerably quantity of pebbles of red beds and green contact-metamorphosed rock, plus a minor amount of Benton shale. These materials contain commercial quantities of gold.

Michigan and Jefferson outwash aprons from Wisconsin stage of glaciation.—The Michigan outwash apron is almost entirely covered by alluvium from Antelope Gulch and other gulches but the Jefferson outwash apron is well-exposed. The most important features of both are the abundance of pre-Cambrian rocks, including granitized schist, and the absence of green contact-metamorphosed rocks.

Gravel along east bank of Michigan Creek, north of Fairplay-Denver highway.—Along the east side of Michigan Creek, from the Fairplay-Denver highway northward, are gravels composed of pre-Cambrian granite, dark gneiss, quartzite, Dakota sandstone, porphyry, and, rarely, Benton shale. These gravels were derived from the head of Michigan Creek. Very scarce fragments of green contact-metamorphosed rock may be found at the highway but not farther north; they mark the easternmost limit of materials derived from the Tarryall Basin.

These gravels, which cover the base of the eastern valley wall, doubtless were derived from a former extension of terrace No. 2.

Alluvium.—The channel and flood-plain materials of present streams are mapped as aluvium as is also the Recent mantle of other depressions. In many places the alluvium grades imperceptibly into outwash aprons, residual mantles of very gentle slopes, and other unconsolidated deposits. At many, though by no means all, such places the topographic boundary drawn on the map coincides with the boundary between cultivated and pasture land, for hay fields, or "meadows," are commonly restricted to areas of recent alluvium.

INTERIOR DRAINAGE BASIN SOUTH OF PARK GULCH

In T. 9 S. and extending from R. 75 W. to R. 76 W. is an area of interior drainage that extends beyond the area mapped. The original valley was excavated by headward erosion from Tarryall drainage, and it once had a smooth pediment floor sloping down to the Tarryall terrace No. 2. While the terrace elsewhere was being dissected by stream erosion, however, this area developed its present form.

The basin consists of two parts. The larger, western part is entirely surrounded by bedrock, locally covered by terrace gravel. The eastern part contains two depressions surrounded by bedrock and, therefore, isolated from the small ravine that flows into Park Gulch. Most of this basin cannot be attributed to stream erosion. Perhaps it is due to wind scour.

GOLD PLACERS

GENERAL FEATURES

PRODUCTION AND FINENESS OF GOLD

Placer gold in Park County was discovered in 1859. From then until 1938, inclusive, the value of placer gold mined in the county exceeded $4,000,000.[14] South Platte Valley has accounted for approximately three-fifths of this value, Tarryall something under two-fifths, and Beaver virtually all the remainder. Half the aggregate

[14] Output of placers during 1939–42 aggregated approximately $2,000,000.

yield occurred prior to 1868, but production has continued, with at least a small output each succeeding year except 1918. Greatest activity since 1868 occurred during 1887, 1922–24, 1935–39.

The value of output in the immediate vicinity of Fairplay prior to 1874 was estimated by R. W. Raymond, as quoted by Henderson,[15] at about $1,000,000. Most of this came from the South Platte Valley, and a subordinate part from Beaver Creek. Subsequent output from Park County until 1904 cannot be allocated to districts. From 1904 to 1938, the output from Fairplay, according to the United States Bureau of Mines figures furnished by C. W. Henderson, amounted to 25,934 ounces, valued at about $500,000. The yearly average fineness of the gold mined since 1904 has ranged from 0.705 to 0.828, with a mean of approximately 0.800.

The Alma placer apparently was started about 1870, and its output was valued at $19,000 during the first 3 years. During the period 1904–38 the yield amounted to 11,989 ounces, valued at $350,000. The yearly average fineness of gold since 1904 has ranged from 0.823 to 0.830.

The Snowstorm placer at the MacConnell Ranch probably has yielded less than the Alma placer. No record of its output prior to 1904 is available, and since 1915 the United States Bureau of Mines has included the relatively small output from Snowstorm with that from Fairplay. During the period 1904–14 the output amounted to 6,390 ounces; a peak was attained from 1909 to 1911. The average fineness of gold during a 5-year period ranged from 0.819 to 0.824.

From the South Platte Valley above Alma 1.97 ounces of gold was produced from 1904 to 1938; the yearly average fineness of this gold ranged from 0.762 to 0.884. The output from Buckskin Gulch amounted to 61 ounces between 1904 and 1938, the average fineness of the gold ranging from 0.791 to 0.826.

Tarryall Creek and its tributaries were credited by R. W. Raymond, as quoted by Henderson,[16] with an output of about $1,000,000 between 1859 and 1872. According to records of the United States Bureau of Mines, furnished by C. W. Henderson, the output during 1904–38 amounted to 5,701 ounces, valued at about $160,000. The total value of output must have been approximately 1¼ million dollars. This amount cannot be precisely allocated among individual placers. Probably the Fortune and Peabody placers together account for more than a million dollars worth of gold, the Wilson for about one-tenth of a million, the placers in upper Deadwood Gulch for somewhat less, and other small placers for the remainder.

---

[15] Henderson, C. W., Mining in Colorado, a history of discovery, development, and production: U. S. Geol. Survey Prof. Paper 138, p. 187, 1926.

[16] Henderson, C. W., idem.

Yearly average fineness of gold from the Fortune placer during 1914–18 ranged from 0.884 to 0.890 and from the Wilson placer during 1934–36 from 0.922 to 0.930. The lowest fineness recorded from the Tarryall district was in 1921, when 12 ounces averaged 0.803; and the highest was in 1912, when 13 ounces averaged 0.967. All other records, exclusive of those of the Wilson placer, show yearly averages ranging from 0.858 to 0.890.

Henderson [17] quotes R. W. Raymond as stating that "Beaver Creek was in the early days a noted creek, but owing to high cost of working it has lain idle for many years." Its output from 1904 to 1938 amounted to 1,538 ounces worth about $40,000, or about one-fifth the output from Tarryall Creek during the same period. All but an insignificant part of the output from Beaver Creek has come from the lower part, where it emerges from the mountains. Yearly average fineness of gold from this locality ranged during 5 years from 0.820 to 0.837. Output from the upper part of Beaver Creek, probably aggregating not much more than 100 ounces, had a yearly average fineness ranging from 0.889 to 0.910.

It may be noted that United States Bureau of Mines records for yearly average fineness show aggregate gold and silver in excess of 0.990 from all except a very few small placers; hence, no metals other than silver are alloyed in appreciable quantities with the placer gold.

The foregoing information and inference have been used to compile the following table, in which the most important placers are arranged according to supposed order of output:

Estimated output of placers in Park County, Colo., to the end of 1938

Name of placer	Location	Estimated output to end of 1938	Range in yearly average fineness of gold
Fairplay	South Platte	$1,900,000±	0.705 to 0.828
Combined Fortune and Peabody	Tarryall	1,300,000±	.860± to .890
Alma	South Platte	400,000+	.823− to .830
Snowstorm (at MacConnell ranch)	do	100,000+	.819 to .824
Wilson	Lower Tarryall	100,000+	.922 to .930
Lower Beaver Creek	Lower part, Beaver Creek	(1)	.820 to .837
Cline Bench	Tarryall	(1)	(1)
Upper Deadwood	do	(1)	(1)

1 Unknown.

DISTRIBUTION

Distribution of gold placers is shown on plate 10. In general, placers located within the maximum limits of glaciers of the last or Wisconsin stage are associated with moraines deposited by ice, whereas those farther downstream are associated with water-laid deposits.

Moraines are recognized by their distinctive hummocky forms and by the sedimentologic character of their materials, which form a

17 Henderson, C. W., op. cit., p. 190.

heterogenous assemblage of rather angular rocks unsorted as to size. By far the most productive localities have been the outer parts of terminal moraines marking maximum ice advance during the Wisconsin glaciation. Those in South Platte and Tarryall Valleys have already been extensively worked and so account for a large share of the output of this region. Smaller though moderately productive placers have been found locally upstream. Some are in secondary terminal moraines deposited during major ice stands or readvances. Those associated with the Alma substage rank second in output. In Deadwood Gulch minor placers are associated with the Briscoe substage. Trivial output here and there has come from other secondary moraines. Still other placers have been worked in lateral moraines; the most valuable have been those in lateral moraine deposited between the time of maximum ice advance and the Briscoe substage. Where bedrock depressions underlie the surface, particularly in coincidence with ice-border channels, the content of gold is likely to be higher than elsewhere. Even lateral moraines deposited subsequent to the Briscoe substage have been worked locally.

Invariably, the bulk of the gold contained in moraines lies adjacent to bedrock. Commonly, two other pay streaks, each about a foot thick, are found above the bedrock, but they are less rich and much more lenticular than the one on bedrock. Rarely have more than three pay streaks been found.

Pre-Wisconsin moraines, which doubtless contained as much gold as the later ones, have not been worked, presumably because only the lateral debris is now exposed. The far more favorable terminal lobes have been either completely eroded or else partly eroded and covered by later outwash.

Outwash aprons which extend downstream from the outermost Wisconsin moraines account for a preponderance of the past output and offer greatest hope for the future. In the South Platte and Tarryall Valleys, production prior to 1939 was confined largely to stream channels, where perhaps there was a slight reconcentration of gold from the surface materials. The bulk of the gold, however, occurs at the base of the outwash materials, adjacent either to bedrock or to older gravel, where it cannot have been appreciably affected by postglacial erosion. Consequently, there is no good reason to believe that the present stream channels are richer than other parts of the apron. In Beaver Creek, by contrast, the entire apron has been worked within the limit of productive sediments.

Outwash aprons are composed of the same materials as the moraines immediately upstream. Their boulders are slightly flatter and rounder, however, and extremely large ones are absent. In spite of their deposition by water, outwash aprons show no stratification and surprisingly little obvious evidence of sorting, but they commonly are shingled. There is a gradual diminution in size of boulders and in content of gold away from the moraine. Gold is not uniformly distributed over the bedrock but occurs in definite channels that must be located by prospecting. Commonly they are associated with channels in the bedrock surface.

The pre-Wisconsin outwash apron of Tarryall Gulch has yielded considerable gold at the Cline Bench placer, 3 miles downstream from the Wisconsin moraine. An outlier of the same apron, to the north, is, therefore worthy of prospecting. Corresponding outwash aprons are absent in South Platte and Beaver Valleys unless they are buried beneath later outwash.

A moderately productive group of placers lies more than 6 miles below any former glacier in Park Gulch and Tarryall Creek. As interpreted in a later section, their gold represents a reconcentration of thinly disseminated gold from the terrace No. 2 gravel. Other placers containing gold from the same source, therefore, may exist where terrace No. 2 has been dissected. In the South Platte Valley, on the contrary, any gold derived from dissection of terrace No. 2 has become mingled with gold in the outwash apron.

Terrace No. 2 gravels, except where dissected and this gold reconcentrated in channels, have not yielded placers and are not favorable for prospecting. Those of Tarryall Creek show small quantities of gold wherever prospected, yet the amount is entirely too small for commercial production. Terrace No. 2 of Crooked Creek and of Fourmile Creek undoubtedly contain much less. For the same reason, gravels of terrace No. 1 are not worth prospecting.

SOURCES OF GOLD

The two major sources of placer gold have been (1) a northeastward-trending mineralized belt in the Mosquito Range, comprising several areas of closely grouped lodes as well as scattered lodes elsewhere, and (2) the Montgomery-Deadwood mineralized area. (See pl. 8). Two independent clues, namely, the fineness of the gold and the lithologic identity of associated materials, determine the source of gold at any specific locality.

Gold from the Montgomery-Deadwood area has had a fineness consistently exceeding 0.858, except for one shipment of 12 ounces during 1921. Its associated materials contain conspicuous quantities of green contact-metamorphosed rock and completely lack pre-Cambrian injection gneiss, schist, and white granite. Pink granite of pre-Cambrian age and pegmatite likewise are entirely absent except where debris derived in situ from the eastern range of granite hills has been admixed. Except for several insignificant placers of Selkirk Creek, the gold of Tarryall Creek and all its tributaries has been derived from the Montgomery-Deadwood area. To the same source may be attributed the gold-bearing though nonproductive gravels of the Tarryall terrace No. 2 and the Park Gulch terrace No. 3. Likewise, this area has supplied all the gold above the Grebb placer in Beaver Creek, although there the quantity of green contact-metamorphosed rock is relatively small. Conclusions based on fineness of gold and lithology conform to physiographic evidence. The bulk of the mineralized area is drained by Montgomery and Middle Tarryall Creeks, so placers are abundant in these gulches as well as in Tarryall Creek below the point at which the Middle Tarryall Glacier spilled over its eastern divide. Upper North Tarryall Creek and Selkirk Creek, in contrast, are virtually barren. Beaver Creek drains only the west tip of the mineralized area, so production from it above the moraine derived from the South Platte Glacier has been slight. As Trout Creek drains the extreme south tip of the area, its gravels doubtless are gold-bearing. Crooked Creek received little or no gold.

Gold from the mineralized belt of the Mosquito Range shows a fineness of less than 0.837, except for one shipment of 16 ounces, averaging 0.884, which was produced in 1917 from workings at the foot of Hoosier Pass. The associated materials are rich in pre-Cambrian rocks and devoid of green contact-metamorphosed rock. All placers of the South Platte and its western tributaries, therefore, and also of Beaver Creek below the Grebb placer derived their gold from this source. This conclusion is confirmed by physiographic evidence, for the mineralized belt is drained by the South Platte River and its western tributaries but not by Beaver Creek. Two miles north of Fairplay, however, ice from the South Platte Glacier spilled into Beaver Creek, and below this point are placers containing gold derived from the South Platte.

It is of considerable interest to speculate regarding contributions from specific areas within the mineralized belt of the Mosquito Range. The North Star Mountain area clearly contributed the bulk of the gold in the Alma placer as well as in very minor placers farther north. Probably it was also the main source of gold in the Snowstorm and lower Beaver Creek placers, for in each the materials came from near the east border of the South Platte ice sheet, and all the gravel between the Alma and Snowstorm placers contains some gold. This distribution of the gold illustrates the tendency of ice and debris to maintain their continuity downstream and not to mingle after glaciers coalesce. Furthermore, the fineness of the gold in the Alma, Snowstorm, and lower Beaver Creek placers shows an almost identical and very narrow range, from 0.819 to 0.837.

The Fairplay placer had no output from 1904 to 1915, 1916 to 1921, and 1928 to 1931. Average fineness of gold during the remaining years has ranged from 0.705 to 0.828, but these figures include whatever output came from the Snowstorm placer and give no clue as to whether consistent geographic variations exist. Dredging operations are reported separately, however, and they are known to have extended from the Fairplay moraine to the south limit of operations. Fineness of gold from the dredging, which includes more than nine-tenths of the total output since 1915, showed averages during four successive years as follows: 0.744, unknown, 0.808, and 0.822. These figures indicate a progressive increase in fineness as the dredged worked southward and downstream, yet at a rate considerably more rapid than normal for placers in general. Of particular interest is the figure for the first year, which presumably represents gold closest to the moraine; it is much below all figures for gold fineness in the Alma, Snowstorm, and lower Beaver Creek placers. A main source other than North Star Mountain, and obviously in Buckskin or Mosquito Gulch, is suggested therefore though by no means proved. The greater fineness of gold recovered during the last 2 years of dredging suggests that part of the gold may have come from a source other than the surface outwash apron—perhaps an underlying, partly eroded moraine and outwash apron of an older glacial stage.

In summary, the Montgomery-Deadwood area and the North Star Mountain area were very definitely major contributors of placer gold; in addition, the Lower Loveland Mountain area, whose northern part yielded the trivial amounts produced from Buckskin Gulch, may have been the main source of gold in the Fairplay placer. Neither the London Mountain area nor the North Star Mountain area, however, can be definitely eliminated as a possible source of the Fairplay gold. The rather strong possibility that Buckskin or Mosquito Gulch is the main source, however, enhances prospects in some of the undeveloped areas upstream.

The relative fineness of gold attributed to each of the sources conforms to theoretical expectations. The highest-temperature deposits, in the Montgomery-Deadwood area, have yielded gold of greatest fineness; intermediate-temperature deposits at North Star Mountain have yielded gold next in fineness; and somewhat cooler-temperature deposits of Lower Loveland Mountain or of London Mountain may have yielded the gold of least fineness.

SOUTH PLATTE DRAINAGE AREA

BUCKSKIN GULCH

Gravel pits testify to former placer operations in Buckskin Gulch at short intervals, extending from the middle of sec. 3, T. 9 S., R. 78 W., almost to Alma. The principal localities are shown on plate 10. During 1939 one man working at the uppermost locality had not obtained enough gold for even one shipment. At the same time, two men working a fifth of a mile farther downstream were obtaining a little gold, but most of it came from the dump of the Orphan Boy mine. From the remainder of the placers small commercial quantities of gold evidently were obtained.

The distribution of placers is closely related to ice stands of the Wisconsin glaciation. During the Alma substage the ice did not reach South Platte Valley. The position of the outer border is uncertain, for the terminal moraine lies against slightly older moraine with no clear-cut separation. The line drawn on plate 9 cannot be far wrong, however, for the inner side of the terminal moraine, where it opens into an interior basin, is clearly defined. From a point a mile upstream from the moraine of the Alma substage where it overlies a rock step formed of resistant Cambrian quartzite, a moraine covers the valley floor for nearly a mile to the Paris Mill where it opens into another interior basin. This moraine shows obscure evidence of at least two ice stands, as indicated on plate 9, and is probably correlated with all four of the moraines just north of Alma.

Most of the placer gold has come from the present stream channel where it cuts through the moraine of the Alma substage, and immediately downstream. There, during postglacial time, the gold probably has been somewhat reworked from the moraine. Next in size is a placer along the left bank of Buckskin Creek in the extreme southwest corner of sec. 2, T. 9 S., R. 78 W. There the gold is in gravel not yet dissected by the stream. It is derived either from materials deposited on the floor of the former lake basin inside the moraine of the Alma substage, or, more probably, from outwash during the next ice stand.

No large output from Buckskin Gulch is to be anticipated, but the digging of a few prospect pits to bedrock near the stream and just inside the margin of the moraine of the Alma substage is perhaps warranted.

MOSQUITO GULCH

All placer gold in Mosquito Gulch, as in Buckskin Gulch, is closely associated with moraines of Wisconsin age. Ice from Mosquito Gulch was tributary to the South Platte Glacier during the Fairplay and Briscoe substages; so these substages are represented only by lateral moraines. The terminal moraine of the Alma substage forms a low but conspicuous belt of hummocks crossing the valley immediately above the mouth of Mosquito Creek and joining lateral lobes on each side. A narrow, inner arc of hummocky moraine (see pl. 9), separated from the outer morainal arc by just a suggestion of an interior basin, may represent either a phase of the Alma substage that cannot be distinguished in South Platte Valley or the equivalent of the first terminal arc north of Alma. The inner arc in turn opens into a well-defined basin extending 1½ miles upstream to Park City.

The open basin terminates abruptly, in the southeast part of sec. 10, T. 9 S., R. 78 W., against moraine that irregularly covers the valley floor for approximately 4,000 feet upstream and confines the stream to a narrow, tortuous gorge less than 50 feet deep. The moraine seems to overlie a rock dam formed by the relatively resistant pre-Pennsylvanian sedimetary sequence. Though obviously not high enough to stop the flow of a strong glacier, the dam may well have terminated ice that otherwise would have extended a short distance farther down the valley, and so the moraine may represent terminal debris from several readvances corresponding to the separate arcs just north of Alma. One ice stand is strongly suggested by coverging ice-border channels in sec. 15.

The only placer worthy of note in Mosquito Gulch is that of the Orphan Boy Placer Gold Mining Co., southeast of Park City. (See pl. 10.) It was operated for more than a month during 1939 but presumably did not meet expenses. Gold was obtained from moraine of post-Alma age from outwash materials.

On the eastern spur of Bald Hill occurs the Dyer placer, a narrow cut more than a hundred feet long. Output from it must have been small. Gold, apparently obtained just above bedrock from lateral moraine lying above the Briscoe ice level, came from Mosquito Creek. There is no obvious reason for a greater concentration of gold in the Dyer cut than elsewhere in the same moraine; perhaps irregularities in the bedrock surface had some influence.

Presence of even small amounts of gold at the two localities suggests that the terminal moraine of the Alma substage is worthy of prospecting. Depth to bedrock in the moraine is unknown but could be determined by a small amount of geophysical work. Gold, whether in commercial quantities or not, could have been derived from either the Lower Loveland Mountain or the London Mountain areas.

SACRAMENTO GULCH

Sacramento Gulch contains no productive placers. Traces of gold in moraine are suggested by several small prospects (see pl. 10) where the inner border of the lowermost moraine opens upstream into an interior basin. Sacramento Gulch drains the north tip of the Sheep Mountain mineralized area, where the lode deposits contain silver but very little gold; no commercial gold has been won either from prospects in "Blue limestone" near the head of Little Sacramento Gulch or from deposits along the London fault in Big Sacramento Gulch. The absence of gold lodes within the drainage area, together with absence of any successful prospects in moraines or gravels, makes Sacramento Gulch a totally unpromising place to prospect.

BEAVER CREEK

Upper Beaver Creek.—Two small placers, in sec. 32, T. 8 S., R. 77 W., occur within the lowermost moraine of the Beaver Glacier, and a third occurs barely outside it. Gold lay on bedrock beneath an unconsolidated mantle only a few feet thick.

The largest placer of upper Beaver Creek is the Miller and Sheldon, half a mile downstream from the terminus of the moraine. Bedrock, chiefly porphyry, is covered by a thin mantle of outwash gravels mixed with talus from adjacent slopes. Gold occurs on bedrock and within shallow unconsolidated fillings of fissures in bedrock. Although appreciable quantities of gold have been extracted, and although a few small rich pockets are found, in general, the yield is meager.

South of the Miller and Sheldon diggings occurs first a small prospect, and beyond it the Carey placer, which is in general similar to the Miller and Sheldon, except that its workings are smaller and its gold content leaner.

Deposits of commercial value are not to be expected south of the Carey placer and north of the moraine derived from South Platte Valley. Between the Carey and the Miller and Sheldon placers small accumulations may possibly be found, but the results to be expected would not justify the expense of systematic prospecting. The small area between the Miller and Sheldon placer and the Beaver moraine could contain some accumulation of gold, yet the most optimistic expectation of finding gold is little better than that of the Miller and Sheldon. Above the terminal moraine, no placers are to be anticipated.

Lower Beaver Creek.—In lower Beaver Creek the Grebb is the northernmost placer in moraine derived from South Platte Valley. At the base of the valley wall are several short, narrow cuts, in which a few pre-Cambrian boulders indicate a pre-Wisconsin moraine. A thousand feet to the southwest, somewhat larger workings in the Wisconsin moraine occur along the bed of a wide, dry ravine called Poorman's Gulch. Three-fifths of a mile still farther south is another small placer in the Wisconsin moraine. A mile and a half north of Fairplay are three small workings; the principal one consists of several narrow cuts, a hundred or more feet long, in moraine, but the other two are mere prospects in talus or in the gravel bed of a former stream east of the creek.

Well up the hillside west of Beaver Creek, mainly near the east border of sec. 29, T. 9 S., R. 77 W., are two placers of moderate size, yet greatly surpassing in importance any previously mentioned. They are a few hundred feet apart. The northwestern pit is roughly 500 feet long, more than 100 feet in maximum width, and about 50 feet in maximum depth; it coincides with a bedrock depression, which doubtless localized the detrital gold. The southeastern pit, approximately 600 feet long and 200 feet in maximum width, apparently follows a bedrock bench, characterized by small hummocks and depressions, along the west wall of Beaver Creek Valley. In both pits the gold lay in moraine just above bedrock. Whether the moraine is of Wisconsin or pre-Wisconsin age is undetermined.

By far the most productive placer in lower Beaver Creek Valley is along the valley bottom, beginning in sec. 29, T. 9 S., R. 77 W., and extending with only two short interruptions nearly a mile downstream. The northern stretch has been worked intermittently since the earliest days of the district. Not until 1938, however, did the Timberline Dredging Co. place a dredge at the south limit of the productive area. By September 1939 the dredge had moved nearly 2,000 feet upstream; it was planned to continue as far as the dredge could go. Value of output during 1939–42 was approximately $350,000. Gold, lying mainly just above bedrock, occurs in outwash gravel essentially similar to that of South Platte Valley. This productive gravel extends from the tip of the former Wisconsin ice sheet downstream to the point where the valley opens into a broad fan, beyond which the gold content decreases abruptly. From plate 10, it will be seen that most of the placer lies adjacent to "older" moraine. The productive gravel is assigned a Wisconsin age, however, on the inference that interglacial erosion removed all older unconsolidated material from the narrow stream bed, whereas the much shorter postglacial erosion has not yet removed the Wisconsin outwash. Present depths to bedrock, nowhere more than 15 feet in the productive area, can reasonably be ascribed to deposition during Wisconsin time.

According to Mr. E. J. Cunningham, of Denver, Colo., dredging was begun at the south limit of productive terrain as determined by prospecting; nearly all the most favorable part of Beaver Creek has therefore been worked. The entire west wall of the valley from its mouth to the north side of Poorman's Gulch, in sec. 17, T. 9 S., R. 77 W., must be regarded as having some prospective interest, however, for the moraine may contain local pockets in addition to those already exhausted. All this ground lies within the Snowstorm property, which has been prospected to some extent, but no information concerning the gold content of the gravel could be obtained. Depths to bedrock in drill holes and pits of the Poorman's Gulch portion, according to the results of drilling and test-pitting furnished by Mr. E. J. Cunningham, are shown together with approximate surface altitudes on plate 11.

CROOKED CREEK

None of the branches of Crooked Creek heads within a mineralized area, so on placer deposits are to be expected in the Crooked Creek area.

SOUTH PLATTE RIVER

Area north of Alma.—According to several operators in Alma, the quantity of placer gold decreases abruptly north of the Alma moraine. Their reports seem to be corroborated by an absence of large placers.

The most important placer upstream from Alma is in the last secondary terminal moraine, located at the foot of Hoosier Pass, near a former town called Montgomery. Gold in small quantities occurs at the base of detrital material, ranging from 1 foot to several feet in thickness, which overlies and fills cracks in the bedrock. Several prospects occur along the outer slope of an ice-border channel marginal to the terminal moraine.

In an open basin that extends 2½ miles downstream from the latest secondary moraine are three very small placers, located about a mile from the upper end of the basin. All are on a low moraine-covered bench that represents the valley floor prior to postglacial erosion; the southernmost is in a channel marking a pause in ice retreat. Although traces of gold can be found, there is little probability of finding placers of commercial value within this basin.

Immediately north of Alma are four rather closely spaced terminal moraines. (See pl. 9.) Each is a hummocky ridge which fills the valley except for a narrow stream gorge, curves outward into lateral lobes bordered by well-defined ice-border channels, and opens upstream into an interior basin. The northernmost moraine, about 2 miles north of Alma, is the weakest and the southernmost, three-quarters of a mile north of town, is the most prominent. Prospecting to bedrock in the outermost part of each moraine may reveal commercially recoverable accumulations of gold, but none is likely to be nearly as productive as the Alma placer.

Alma placer.—The Alma placer lies entirely within the moraine of the Alma substage. This moraine, unlike the Fairplay and Briscoe moraines, is not hummocky, nor does it possess terminal or lateral ridges. It is a low, smoothly rounded hill, nearly a mile long and parallel with the valley.

The placer extends for nearly a mile along the east bank of the South Platte River. (See pl. 16, *A*.) There are two main pay channels, trending approximately parallel with the stream and separated by a low ridge that indicates bedrock at shallow depth. Thus, the eastern channel follows a bedrock and topographic trough that separates the valley wall from the low dividing ridge, whereas the western channel lies near the crest of a steep slope bordering the river. At places the two channels extend across the dividing ridge and coalesce. According to one local resident, the channels average between 40 and 50 feet in width, but according to another they are only 12 to 20 feet in width.

In detail, the picture is far less simple. The bedrock surface is not smoothly curving but extremely irregular, with many small hollows, troughs, hummocks, and ridges. High concentrations of gold values, according to testimony of several operators, were not necessarily confined to the local depressions or always in them. Many pay streaks lie along the slopes and a few actually lie on the crests of hummocks; nevertheless, in general depressions have been more productive than ridges. Unfortunately, sufficient information is not now available either for making a detailed contour map of the bedrock or for outlining the richest terrain.

The gravel contains slightly rounded boulders characteristic of glacial debris. They show little evidence either of sorting or of stratification, except at a few places. Few boulders are of the huge size seen elsewhere in the moraines. (See pl. 16, *B*.) In a preliminary study of the fine materials, Miss Ethel Davis, of Rochester, N. Y., found that quartz and feldspar predominate, that magnetite is the most abundant of the heavy constituents, and that garnet chlorite, epidote, and other minerals occur in minor quantities.

The richest and most continuous pay streak is less than a foot thick and lies immediately above bedrock. According to Mr. McLoughlin, the gold in it is about 0.025 finer than gold in overlying streaks. Near the north end of the Alma placer, two other pay streaks, each about a foot thick, occur approximately 10 and 25 feet, respectively, above bedrock. Each is lenticular over short distances and is slightly more stained by limonite than the material above and below. In the remainder of the placer, one or both of the upper streaks occur sporadically, and there are traces of gold throughout the gravels.

There has been no adequate prospecting either in the river channel or along the west bank although both merit investigation. The river bed, particularly the area adjacent to the lower part of the Alma placer, may contain gold washed down the slopes or reconcentrated from eroded portions of the moraine during postglacial time. Depth or location of any bedrock channels beneath the flood plain of the river is unknown but can readily be determined by geophysical work.

Van Epps placer.—The Van Epps is a small placer along the bed of a ravine in the western part of sec. 7, T. 9 S., R. 77 W. Gold occurs in lateral moraine deposited prior to the Briscoe substage. There is no apparent cause for concentrations of gold at this locality but the ravine may overlie a bedrock depression, or there may have been a slight reconcentration of the gold by recent stream action.

Northern part of Snowstorm property.—What is here regarded as the northern part of the Snowstorm placer property extends along the east side of the South Platte River from the Alma placer nearly to the MacConnell ranch. This area includes all the terrain from the outer margin of the Briscoe substage to the outer margin of the Alma substage. (See pl. 9.)

A record of depth to bedrock in bore holes and pits over part of the terrain has been furnished by Mr. E. J. Cunningham, of Denver, Colo. These depths, together with approximate surface altitudes, are shown on plate 11. This map includes parts of all three topographic features, namely the Briscoe moraine, the ridge that trends parallel with the valley upstream from the Briscoe moraine, and the benchlike depression between them. It will be seen that the moraine cover ranges in thickness from 15 to 55 feet. Farther north the thickness presumably is comparable to that in the existing prospect holes.

No precise information concerning gold content was obtained. Current reports indicate that at least traces may be found everywhere; moreover, several small placers were being operated by lessees during 1939, and larger ones (not on pl. 10) were opened in 1940. The ground must therefore be regarded as favorable for prospecting.

Small operations seen by the writer reveal that the greatest gold concentration immediately overlies bedrock. In places there is a 1-foot layer of cemented gravel of unknown age beneath the unconsolidated mantle.

Snowstorm placer at MacConnell ranch.—The original Snowstorm placer was worked along two cuts, an older one just north of the MacConnell ranch and a newer one less than half a mile to the south. They are located in secs. 19, 20, 29, and 30, T. 9 S., R. 77 W., and are shown on plates 10 and 11. Gold occurs in lateral moraine not far outside the limits of the moraine of the Briscoe substage.

The newer cut heads at the top of a morainal bench that probably represents the first ice stand (see pl. 9) above the Fairplay moraine. For a brief distance it crosses the bench, then curves to follow an ice-border channel marking the second glacial stand. Two short arms branch westward from the main cut. Configuration of bedrock in and around the cut could not be determined, but, as at Alma, there is evidence of local surface irregularities. Placering extended to fairly shallow bedrock, in general less than 30 feet deep, from the head to within several hundred feet of the pit entrance. According to Mr. William McKinley, lessee during 1939, depth to bedrock increases abruptly near the lower end of the cut and attains about 100 feet at the entrance.

A map in possession of State Senator Price Briscoe shows an average gold content of 21 cents per cubic yard in the cut and 16 cents per cubic yard in the unworked pillars. According to 1939 lessees there are three pay streaks in most of the cut. A pink one overlying bedrock, probably derived its color from the red layers in the rock beneath. Two irregular pay streaks lie at various distances above bedrock; the distribution of their gold is very pockety. The gold originally was deposited as part of the lateral moraine, but was probably slightly reconcentrated by channel water and perhaps also by slight additions during the ice stand.

The older cut of the Snowstorm placer lies northwest of a ridge represented by slight bends in contours on plate 10. Prospect pits, as indicated by plate 11, show that the ridge consists of bedrock at shallow depth. This bedrock ridge must have impeded ice flow and concentrated gold in the trough immediately upstream. Placering extended to bedrock and revealed many irregularities.

East side of valley between Fairplay moraine and main Snowstorm cuts.—Despite widespread traces of gold in moraine east of the South Platte River and of commercial production in cuts near the MacConnell ranch, there is no record of adequate prospecting in ground between the main Snowstorm placer and the outer margin of the Fairplay moraine. The surface is covered with lateral moraine, which warrants testing immediately above bedrock. The considerable surface relief doubtless reflects bedrock relief; consequently, surface depressions and channels would seem the most favorable localities in which to begin prospecting. A medium-sized placer (not shown on pl. 10) was started in 1940 on the east side of the Alma-Fairplay highway.

West side of valley between Sacramento and Mosquito Creeks.—Only two very small placers exist west of the South Platte River between Sacramento and Mosquito Creeks. One of them, the Dyer, has already been described in the section on Mosquito Gulch. The other is in sec. 25, T. 9 S., R. 78 W. (see pl. 10); it lies along the bottom of a shallow, narrow ravine, interpreted as an ice-border channel (see pl. 9) equivalent to the one south of the MacConnell ranch. Bedrock has not been reached.

Absence of other placers does not necessarily condemn the ground, for almost no prospecting to bedrock has been done; on the contrary, the presence of gold at these two localities suggests that the lateral moraine may be gold-bearing elsewhere. Chances for finding commercial deposits, however, should be regarded as entirely fortuitous. The reader should keep in mind that gold east of the river has no relation whatever to possibilities west of the river.

The most favorable places to begin prospecting would be in the ice-border channel adjacent to the terminal part of the Briscoe moraine and at the mouths of other ravines in lateral moraine to the south. At such places slight reconcentrations of any original gold may be expected.

River bed between Sacramento and Mosquito Creeks.—The stream and flood-plain channel of the South Platte River above Sacramento Creek has not been tested to bedrock. Depth to bedrock and possibilities for commercial placers, therefore, are not known. The most favorable localities for prospecting seem to be just downstream from commercial placers in the bank.

Fairplay placer.—Extending approximately from Sacramento Creek to more than a mile southeast of Fairplay (see pls. 10 and 17, *A*), is a placer, commonly called the Fairplay placer, that crosses several properties. A great deal of the gold in this placer came from the Fairplay moraine, but considerably more came from outwash materials beyond.

The Fairplay terminal moraine is a sharp-crested, hummocky ridge that crosses the Platte Valley in a crescent-shaped curve. From its outer border, indicated on plate 9, abrupt slopes rise more than 100 feet above the outwash apron to the south; the inner or north side is bounded by a broad morainal bench made up of irregular hummocks and depressions at an average height of slightly less than 50 feet below the ridge top. This bench opens upstream into a huge, roughly circular basin formerly occupied by a lake. The narrow frontal ridge on the northeast side of the South Platte River curves abruptly but extends for a mile as a lateral moraine flanked on its outer side by a conspicuous ice-border channel, then it merges with the Platte-Beaver Divide.

Southwest of the river the frontal ridge extends as an unbroken topographic feature to merge with the divide south of Sacramento Creek; the inner bench, on the contrary, turns northward, crosses Sacramento Creek, and joins the lateral moraine along the west side of the Platte Valley. The Fairplay moraine is transected by a V-shaped gorge, 150 feet deep, cut during and subsequent to ice retreat and occupied by the South Platte River. Several other channels, now dry, are clearly represented by configuration of contours in secs. 31 and 32, T. 9 S., R. 77 W. The channel in which a narrow tongue of glacial outwash is between younger and older moraine in section 32 (see plate 10) was formed when the ice was farthest advanced. The others, however, probably continued to develop even after the ice had retreated to the morainal bench.

Workings within the moraine are chiefly along the sides of the steep gorge cut since the ice retreat, yet locally they penetrate farther back from the stream. Judging from intensity of work done, the bulk of the gold came from within 1,000 feet of the moraine border, and most of the remainder from the next 1,000 feet. Still farther upstream, the relatively small size and the lack of continuity of the placers indicate that gold content progressively decreased. In the southwest bank, away from the river, several small pockets have been worked, as shown on plate 10.

Though they will probably be of lower grade and less extensive than ground already worked, there is good chance of locating additional gold pockets or channels that are directly associated with the Fairplay moraine. Along the northeast side of the South Platte River prospecting immediately above bedrock is warranted both in the outermost 1,000 feet of the moraine itself and in the adjoining ice-border channel continuously from the river to or beyond the point where the terminal arc curves to become a lateral moraine. Whether or not additional productive channels may be discovered along the southwest side of the stream, close to the moraine border in sec. 32, T. 9 S., R. 77 W., involves the question of the quantity of gold deposited west of the present stream; prospecting certainly is warranted. On both sides of the present workings, search should be made immediately above bedrock, which may be much more than a hundred feet below the surface. Should no additional commercial deposits be located after adequate prospecting southwest of the present workings, then the lateral moraine west of the river between Sacramento and Mosquito Creeks must be regarded as rather unpromising.

Placers in the Fairplay moraine continue downstream without interruption into glacial outwash. Work prior to 1939 had been largely confined to the postglacial stream channel and its immediate banks. Only at the extreme southeast end did the dredge turn out of the river bed to cut a third of a mile across the southwest bank.

Location and relations of the placer cut below the moraine are shown on a large scale on plate 12. The gravels contain poorly rounded boulders up to nearly 2 feet in diameter (see pl. 17, *B*), though the maximum size gradually diminishes away from the moraine. About the only evidence of deposition under water is the general parallelism of the flatter fragments. Except for a short distance below the moraine, the cut extends to bedrock, where presumably the greatest gold content was found. In a general way, the gold content decreased progressively downstream. Abandonment of operations at the southeast end, according to local reports, was due partly to a decrease in gold content and partly to controversy with ranchers regarding water pollution.

UNDEVELOPED GROUND SOUTH OF FAIRPLAY MORAINE

In 1939 there remained a large unexplored area within the South Platte Valley south of Fairplay, although considerable testing already had been done west of the river. Plate 12, with accompanying cross sections, shows physiographic features in the most favorable part. It also gives the surface altitude and depth of holes for which

A ALMA PLACER.

View from north end of town looking southeast toward northern part of placer workings.

B ALMA PLACER.

Close-up of productive gravel. Note absence of sorting or stratification as well as absence of very large boulders.

A FAIRPLAY PLACER.

View from north end of workings, in moraine, looking southward into South Park. Hogback of Dakota sandstone in distance.

B FAIRPLAY PLACER.

View at south end of workings, in glacial outwash apron, where dredge was abandoned in 1925.

records were furnished by Mr. R. W. Derby, Jr., of La Grange, Calif. No information about gold content was obtained, however, except Derby's statement that it seems to justify commercial development. Prospecting continued during 1940 and led to installation of a dredge on the southwest bank of the river, about a mile from Fairplay, and large-scale commercial operation began in 1941. In general, the gold content may be expected to diminish gradually southward, though a locally rich area may be found about a mile southeast of Fairplay, as the surface gravel there probably overlies an uneroded remnant of older moraine and outwash. Prospecting should be extended downstream until the south limit of commercial production has been defined.

East of the river, and extending to beyond the Purcell ranch, is a plain composed of outwash material but standing about 9 feet topographically lower than the apron west of the river. This difference in altitude is due to postglacial erosion. Because gold would originally have been deposited mainly at the base of the gravel, the postglacial erosion has in no way affected the occurrence of placer deposits. This ground, therefore, warrants prospecting; if commercial quantities of gold are found, testing may be continued southward to the Crooked Creek confluence. Because the surface gravel derived from glacial outwash of the South Platte may possibly overlie a layer of preglacial gravel derived from Crooked Creek, the maximum gold content in this area will not necessarily be found immediately above bedrock.

FOURMILE CREEK

The Hall placer (sec. 16, T. 10 S., R. 77 W.) is the only one worked within the drainage area of Fourmile Creek. Three cuts, the two largest of which are joined, have been made to depths ranging from 5 to 8 feet on an eroded tip of terrace No. 2 a third of a mile west of the Hall Ranch. Their size suggests a minor commercial yield.

The gold of this placer occurs in surface gravel residual from terrace No. 2 and overlies a "false bedrock" of stratified silt, sand, and fine gravel. The productive gravel contains few boulders more than a foot in diameter. All the pebbles, cobbles, and boulders are fairly well rounded and somewhat flat. A few of the boulders are of pre-Cambrian rock. The gold and also the boulders of pre-Cambrian rock probably were derived originally from the South Platte Valley.

On September 14, 1938, a 3.2-ounce shipment of gold having a fineness of 0.831 was made from screenings in a cut on the west side of the Fairplay-Buena Vista highway, 2 or 3 miles south of the Nelson ranch. The locality is south of the area shown on plate 10.

In spite of gold showings at these two localities, the drainage area of Fourmile Creek is not favorable for prospecting, because even from the most optimistic viewpoint results are not likely to pay expenses.

TARRYALL DRAINAGE AREA

MIDDLE TARRYALL VALLEY

Montgomery Gulch.—Placers in the upper part of Montgomery Gulch and its tributary, Purgatory Gulch, have little commercial significance. Several men working in Purgatory Gulch found traces of gold but not in commercial quantity during the summer of 1939. Below the mouth of Purgatory Gulch, upturned morainal gravels in the stream bed reveal the larger former workings in upper Montgomery Gulch. Two very small placers, likewise in moraine, in the western part of sec. 15, T. 8 S., R. 77 W., have not yielded a commercial output. An interesting though small placer in the extreme southeast corner of sec. 16, T. 8 S., R. 77 W., reveals small quantities of gold in talus derived nearly in situ. At the lower end of Montgomery Gulch are placer workings that extend into Middle Tarryall Creek, and are therefore described in the next section. Large commercial placers are not to be expected within Montgomery Gulch, in spite of the fact that it probably was an important contributing source of gold found farther downstream. If further prospecting in Montgomery Gulch is attempted, the most desirable place would be just inside the border of the moraine of the Alma substage. (See pl. 9.)

Deadwood Gulch.—Deadwood Gulch has yielded considerably more gold than Montgomery Gulch, for there are a number of workings whose size suggests a noteworthy profit. Placer diggings in Deadwood Gulch are virtually continuous from the juncture of the two headwater branches west of the Ute mine to a point below the juncture with Montgomery Creek. The two localities that have been most productive are the uppermost half mile of the placer ground and the stretch below the moraine of the Alma substage. Gold in the former occurs in glacial moraine and in the latter in outwash gravels formed during the Alma substage. As the most intensely worked portion of the outwash gravel is adjacent to the junction with Montgomery Gulch, it seems probable that gold from Deadwood Gulch was augmented by outwash from Montgomery Gulch during the Alma substage.

The most favorable ground remaining in Deadwood Gulch lies within the area already worked, so no important extensions are to be anticipated; however, there still remains some unworked gravel.

Middle Tarryall Creek.—In sec. 13 and 14, T. 8 S., R. 77 W., there is a continuous series of placer workings within the Ironwood and Hausclaus properties. Although no figures are available, output from the workings was probably much less than from the ground near the juncture of Montgomery and Deadwood Gulches. Work has been done mostly within the stream and along the flood plain, but locally it was extended into the banks; one cut several hundred feet long lies entirely in the north bank. It will be seen from plates 9 and 10 that the placers begin in moraine of the Briscoe substage and extend for a third of a mile downstream. Nearly all the gold, therefore, except in the cut in the north bank, probably was deposited either with moraine or as outwash during the Briscoe substage. If this be true, then the ground between the Ironwood placer and the lower limit of the placer at the Montgomery Creek confluence is rather unpromising.

Several placers occur, as shown on plate 10, associated with an ice-border channel that can be recognized on both sides of Deadwood Creek. Commercially, all are rather insignificant. They indicate, however, that the lateral moraine contains traces of gold, and justify the suggestion that accumulations may be found elsewhere in small pockets of similar size and richness. The cost of prospecting, however, hardly warrants a search.

Two other very small placers may be regarded as outliers of the main Fortune workings.

NORTH TARRYALL VALLEY

The largest placer in North Tarryall Gulch, above the Fortune, is located in the southwest part of sec. 12, T. 8 S., R. 77 W. It is a cut, approximately a thousand feet long, in moraine derived from the Middle Tarryall glacier and covering the west wall of North Tarryall Valley. Four other very small placers, as shown on plate 10, along the right bank of North Tarryall Creek are in the same moraine. Gold in all of these placers was derived from Deadwood Gulch. Undiscovered gold pockets of comparable magnitude may exist elsewhere, but cost of prospecting is likely to make a search for them unprofitable.

Two very small workings, one in Selkirk Gulch, the other in the left bank of North Tarryall Creek indicate that unconsolidated materials outside the limits of the Middle Tarryall Glacier contain traces of gold, but no commercial deposits have been found or are to be expected. The gold must have come from some unimportant source, probably near the head of Selkirk Gulch.

FORTUNE AND PEABODY PLACERS

The Fortune and Peabody properties are adjacent, and the main workings of both form a continuous belt roughly 2 miles long. Together they have produced most of the gold found in the Tarryall district. Recent production has been confined to intermittent small operations by lessees. The Fortune placer lies mainly within the Wisconsin moraine of the Middle Tarryall Glacier, the Peabody placer in outwash gravels downstream.

Just as at Fairplay, nearly all the gold produced from the moraine lay close to the outer margin. Plate 10 shows that none of the main workings extends much more than a thousand feet inside. Most of the gold lay adjacent to an extremely irregular bedrock surface but pay streaks above bedrock in slightly sorted layers interfinger with unsorted, more clayey debris, resulting in a rather erratic distribution. Mr. Edward S. Ames, of Denver, Colo., has kindly furnished information from more than 24 recent prospect pits and holes on the Fortune placer. They show depths to bedrock ranging from 10 to 35 feet over the moraine and gold value ranging from 1 cent to 40 cents per cubic yard. As the pit showing the greatest content was adjacent to a placer cut, the ground already worked may have been considerably richer. As about half the holes show gold content exceeding 10 cents per cubic yard, it is probable that additional parts of the moraine may yet be worked.

Workings below the terminal moraine, mainly on the Peabody property, lie along the stream bed and in both banks. The greatest gold content is in gravel adjacent to bedrock, but all the gravel contains some gold. The productive gravel was first deposited as glacial outwash, then was partly eroded by postglacial stream action. Unlike conditions in the South Platte Valley, the outwash was confined between relatively narrow valley walls from the ice terminus for a mile downstream where it opens into a broad fan.

Materials of the moraine are essentially the same in composition as those of the outwash apron, but boulders in the outwash apron are slightly more rounded, have smaller maximum diameter, and have slightly more nearly parallel arrangement than in the moraine. An abundance of green contact-metamorphosed rock proves that most of the material, including the gold, came from the heads of Montgomery and Deadwood Gulches.

CLINE BENCH PLACER

The Cline Bench placer is located along the right bank of Tarryall Creek, 2½ miles east of the mountains. Its physiographic relations are indicated on plate 10 and these for a smaller area on larger scale on plate 13 (map and profile sections). Figures of output are not available, but no doubt the enterprise was profitable. The output may have been somewhat greater than that from the placer at the juncture of Montgomery and Deadwood Gulches but probably was much less than that of the Fortune and Peabody placers.

The character of the bench is best seen from the profiles. Its bedrock floor rises abruptly from beneath the stream bed, flattens southward, and then again rises toward the preglacial terrace No. 2. Along the north face, bedrock can actually be seen at places in the placer cut and also in exposures in the dissected portion farther west; elsewhere, bedrock has been reached in many prospect pits.

Gravel which overlies the bedrock ranges from 7 to 36 feet in thickness. It forms three layers (see pl. 15, *B*); the middle, finer-grained layer is lenticular and therefore absent from many places. The following section may be seen in a cut near the north end of the placer:

Section near north end of Cline Bench placer

	Feet
Surface soil	1
Pale-yellow gravel containing numerous boulders up to 1 foot in diameter	2–3
Deep-yellow, slightly clayey sand containing only a few small cobbles and pebbles. This layer apparently is very uneven and lenticular	2–3
Gray to pale-yellow gravel containing an abundance of boulders, cobbles, and pebbles up to 1 foot in diameter	6+
Bedrock; yellow sandy shale (Montana age, Upper Cretaceous).	

As seen from the maps, placer workings are largely but not entirely confined to the north face of the bench. Prospecting at localities indicated on plate 13 reveals, according to Mr. Edward S. Ames, of Denver, Colo., gold value ranging from 1.5 cents to 30 cents per cubic yard along the bench. One channel roughly parallel with the present workings but extending much farther east and west seems rich enough to be worked profitably. The bulk of the gold is adjacent to bedrock.

The surface level of the bench if projected westward across the dissected portion seems to join the uneroded outwash apron downstream from pre-Wisconsin moraine; the bench is therefore interpreted as being of that age. The presence of three distinct gravel layers suggests either that channels shifted due to braiding at the time outwash gravels were deposited, or that the lower gold-bearing gravel may be still older. Whatever the origin of the gravel layers the topographic position of the bench shows that both the preliminary erosion and all subsequent deposition took place after dissection of terrace No. 2 had been begun and prior to the last pre-Wisconsin interglacial stage.

An abundance of green contact-metamorphosed rock in the productive gravel proves that the gold came from the mineralized area of Montgomery and Deadwood Gulches.

PLACERS IN PARK GULCH

The North Branch of Park Gulch contains two placers in sec. 1, T. 9 S., R. 76 W. The larger, which is a few feet wide and nearly 2,000 feet long, must have yielded a small profit, but the smaller is hardly more than a prospect. Both are in the bottoms of small ravines that are separated from Tarryall Creek by bedrock. In each, however, gold is associated with gravel containing green contact-metamorphosed rock. At least in the smaller placer, gold overlies a clay ("false bedrock"). The immediate source of both the gold and the gravel can only be eroded portions of terraces in which the ravines are located.

Near the southwest corner of sec. 17, T. 9 S., R. 75 W., is a cut several hundred feet long. Though it lies in the right bank of Park Gulch, and therefore slopes down from the granite hills, the productive gravel contains considerable green contact-metamorphosed rock. The gold, therefore, came originally from the head of Montgomery and Deadwood Gulches. The bedrock, only a few feet below the surface, is granite.

A fourth, insignificant pit is located in the right bank of the middle branch of Park Gulch, in sec. 28, T. 8 S., R. 76 W.

WILSON PLACER

The Wilson placer is the most important producer in Park Gulch. It extends from the middle of sec. 17, T. 9 S., R. 75 W., to Tarryall Creek. According to the foreman, Mr. Axtell, the value of output from 1934 to 1939 amount to $175,000, the gold content ranging from 10 cents to $1 per cubic yard; one rather large block of ground averaged 35 cents per cubic yard.

The gold has a fineness exceeding 0.900 and maintains a fairly uniform size, particles averaging about a millimeter across. It occurs in a coarse gravel layer about 2 feet thick, 6 to 8 feet below the surface. In the southern part of the cut, most of the productive gravel overlies a bluish-green micaceous clay, but in the northern part, where the stream lies closer to the granite hills, the gravel rests on granite.

The gravel contains about equal parts of granite from the hills to the east and of materials from the head of Tarryall Creek. The granite pebbles, cobbles, and boulders are considerably the more angular. The materials from Tarryall Creek consist of Dakota sandstone, subordinate amounts of porphyry and green contact-metamorphosed rock, minor amounts of Benton shale, and still smaller amounts of red beds.

In 1939 plans called for one shovel to extend the workings up the North Branch of Park Gulch; the ground already had been tested as far as the Spindle Ranch, in sec. 7, T. 9 S., R. 75 W. Another shovel, started in May 1939, was intended to work northward in Tarryall Creek over ground not yet prospected.

The gold originally came from the heads of Montgomery and Deadwood Gulches. This source is proved by the fineness of the gold and its association with considerable quantities of green contact-metamorphosed rock, whose presence in Park Gulch upstream from Tarryall Creek must be explained. The theory that an ancient channel connects the Wilson placer with the point where Tarryall Creek emerges from the mountains is untenable. Careful inspection of gopher holes along the steep slopes of terrace No. 2 reveals a bedrock distribution that precludes any former channel except the middle branch of Park Gulch. At the head of this branch, depth to bedrock beneath the pre-Wisconsin outwash apron is not known, so Tarryall Creek could perhaps have drained through there prior to deposition of the apron; however, erosion along the entire middle branch of Park Gulch has been slight as compared with erosion north of Tarryall Creek, so the main drainage could not have been through Park Gulch for any appreciable time. Besides, bedrock walls along the gulch for several miles southeastward from Como show that, even if a thin layer of gold-bearing gravel did exist, it has been removed.

By far the most probable immediate source of the Wilson placer gold and its associated gravels, other than granite, is the eroded part of terrace No. 2. Small amounts of gold, well below commercial quantity, have been found wherever terrace No. 2 gravel has been prospected east of Como. Extensive dissection of this terrace in the lower reaches of Park Gulch doubtless reconcentrated much of the gold against the barrier of granite hills. According to this intepretation, the richest ground has already been worked, but the North Branch would be highly favorable for prospecting and the main branch east of the ridge of Laramie (?) age rocks somewhat less so; as already mentioned, results of tests in the North Branch have encouraged plans for commercial operations. Perhaps even along the west face of the ridge of Laramie (?) age rocks Park Gulch has slight possibilities.

PLACERS OF TARRYALL CREEK WITHIN EASTERN GRANITE HILLS

A discontinuous group of placers extends down Tarryall Creek, along its course through the granite hills of the east border. The principal localities are shown on plate 10, though some minor ones extend beyond the area mapped. Total output must have been considerable, though decidedly less than that from the Wilson placer.

Unconsolidated debris on the floor and banks of the narrow gorge consist of pre-Cambrian granite from the local bedrock mixed with diverse proportions of materials from the head of Tarryall Gulch. Where the two are approximately equal the ground has been worked, but where granite greatly predominates the ground is barren. The upper Tarryall materials consist of porphyry, green contact-metamorphosed rock, red beds, Dakota sandstone, and Benton shale. Fragments are fairly well rounded and do not exceed 4 inches in diameter. Granite debris, on the contrary, is rather angular and includes large boulders.

Clearly, the gold originally came from the heads of Montgomery and Deadwood Gulches. Present distribution, however, is doubtless due primarily to reconcentration of gold disseminated over eroded portions of Terrace No. 2; in other words, the workings form a downstream continuation of the Wilson placer.

UNDEVELOPED AREAS IN SOUTH PARK

Tarryall Creek.—Within the drainage area of Tarryall and Michigan Creeks there remain large areas of terrain having at least some possibilities for production.

The most favorable locality is immediately east of the point where Tarryall Creek emerges from the mountains. Prospecting there by Mr. Edward S. Ames, of Denver, Colo., prior to 1939 had proved several channels containing gold in commercial quantity. In 1941, after this report was written, commercial operations began in this area. The locations of Ames' prospect holes and pits, together with surface altitudes and depths to bedrock are shown on plate 13 and in profiles that accompany it. Gold values ranged from 0.4 cent to 40 cents per cubic yard, and probably more than half the prospects yielded more than 10 cents.

Comparison of plate 10 with plate 13 shows that Ames' prospects cover less than half the favorable terrain. North of them lies an extensive area covered by outwash from the glacier of Wisconsin age and also a large island of earlier outwash correlated with the Cline Bench. Eastward and northeastward, the Cline Bench terrace merges topographically with the later outwash apron; thus, all the ground east of the mountains as far as the southernmost tributary of Michigan Creek may be regarded as having placer possibilities. As, in a general way, the gold content may be expected to decrease gradually eastward, away from the mountains, the western part should be tested first.

Outwash from the last glacier may not be expected to contain gold in commercial quantity for more than a very few miles beyond the ice terminus; however, gold disseminated over terrace No. 2 and even No. 1 may have been reconcentrated into commercial deposits in any of the places where the terrace has been sufficiently dissected. Gold placers therefore, could conceivably be found anywhere along Tarryall Creek. Results of exploration upstream along Tarryall Creek from the Wilson placer will have special bearing on this question. Should gold in commercial quantity be obtained and should it continue for any considerable distance away from the eastern granite hills, then all of Tarryall Valley may be regarded as potentially favorable ground.

In summary, the entire outwash apron for several miles east of the Peabody placer, as well as the Cline bench and its outlier, are favorable for prospecting; in the southern part of this terrain, several promising channels have already been found. Less favorable, yet well worthy of consideration, is the entire valley of Tarryall Creek between the Fairplay-Denver highway and the eastern granite hills; there, gold may have been reconcentrated from eroded terrace No. 2.

South Branch of Park Gulch.—There are no placers along the South Branch of Park Gulch, though prospect holes have been drilled in the vicinity of a ranch in sec. 4, T. 9 S., R. 76 W. Records of these holes are not available.

The South Branch was the site of Trout Creek prior to its capture by a branch of Crooked Creek subsequent to an early glacial stage. A major part of its terrace No. 3 gravel, as well as its alluvium, therefore, was derived from Trout Creek, yet at least a small part came from the dissected Tarryall terrace No. 2. The deposits contain appreciable quantities of green contact-metamorphosed rock, and so are probably gold-bearing. The entire valley, therefore, must be regarded as possible placer ground, though considerably less favorable for prospecting than Tarryall Creek. The areas northwest of the Fairplay-Denver highway or at the western base of the ridge of rocks of Laramie(?) formation are probably the richest.

MICHIGAN CREEK

UPPER MICHIGAN CREEK

The Stoll placer is a small cut at the north tip of terrace No. 3, which forms the divide between Michigan and Jefferson Creeks. The productive gravel at the bottom of the cut is nowhere exposed. Boulders washed from it, however, are more or less angular and range up to more than 2 feet in diameter. Overlying the productive layer and exposed at several places along the sides of the cut is a faintly stratified gravel whose boulders do not exceed 6 inches in diameter. These relations suggest, therefore, the pay streak was in an older, coarse gravel that may be an uneroded remnant of moraine overlain by outwash material of pre-Wisconsin age.

The Stoll placer proves the presence of at least small quantities of gold in materials from Michigan Creek, but the absence of any other workings suggest that commercial accumulations of any magnitude are improbable. If some do exist they are most likely to extend for a short distance downstream from the moraine in sec. 27, T. 7 S., R. 76 W., where an outwash apron is covered by alluvium from Antelope Creek and other small creeks.

LOWER MICHIGAN CREEK

Traces of green contact-metamorphosed rock in the western part of sec. 18, T. 8 S., R. 75 W., in gravel exposed in a highway cut, show that at some former time minor quantities of Tarryall material were carried into Michigan Creek. This process was probably accomplished during early stages of the dissection of terrace No. 2. The quantity of material deposited may have been sufficient to permit gold deposition in alluvium of Michigan Creek east of Michigan station. However, the chances for finding workable deposits are much less favorable than in lower Tarryall Creek.

PENNSYLVANIA MOUNTAIN

The Pennsylvania Mountain placer, located at an altitude of approximately 12,250 feet on the broad, sloping crest of the mountain (see plate 3) is an interesting curiosity. It has yielded only a small output, owing in part at least to difficulty in obtaining water. The gold there occurs in grains showing great variation in size. The largest nugget, obtained in August 1938, weighed 11.12 ounces, and was 3 inches long and 2½ inches in maximum diameter. Many nuggets showed virtually no rounding, and a few were still clinging to quartz. Recorded output during 1935–37 amounted to 32 ounces of gold, ranging in fineness from 0.743 to 0.749.

The gold occurs well above the ice border of the latest Mosquito Glacier and above any probable edges of earlier glaciers; moreover, it is contained in very coarse debris composed of angular fragments of rocks cropping out in the vicinity. All features therefore suggest that the gold has been derived nearly in situ in an eluvial placer.

INDEX

www.ingramcontent.com/pod-product-compliance
Lightning Source LLC
Chambersburg PA
CBHW032014190326
41520CB00007B/473
* 9 7 8 0 9 8 4 3 6 9 8 3 6 *